Wenchuan Hu

Algebraic Cycles and Lawson Homology

Wenchuan Hu

Algebraic Cycles and Lawson Homology

An Application of Homotopy Theoretic Methods in Algebraic Cycles Theory

VDM Verlag Dr. Müller

Imprint

Bibliographic information by the German National Library: The German National Library lists this publication at the German National Bibliography; detailed bibliographic information is available on the Internet at http://dnb.d-nb.de.

Cover image: www.purestockx.com

Publisher:
VDM Verlag Dr. Müller Aktiengesellschaft & Co. KG , Dudweiler Landstr. 125 a, 66123 Saarbrücken, Germany,
Phone +49 681 9100-698, Fax +49 681 9100-988,
Email: info@vdm-verlag.de

Zugl.: Stony Brook, Stony Brook University, Diss. 2006

Produced in USA and UK by:
Lightning Source Inc., La Vergne, Tennessee, USA
Lightning Source UK Ltd., Milton Keynes, UK
BookSurge LLC, 5341 Dorchester Road, Suite 16, North Charleston, SC 29418, USA

ISBN: 978-3-639-01886-8

Acknowledgements

I am indebted to many for their support and encouragement.

First of all, I would like to express my deepest gratitude to my adviser, Prof. Blaine Lawson, for his insight, encouragement and outstanding guidance. He has been a remarkable influence since I started working with him. His advice, support and patience with me have been critical during my studying at Stony Brook. His help is beyond mathematics. I would have been nowhere without them.

I would like to express my special gratitude to my wife for for constantly standing beside me and for keeping alive the place that I will always call *home*. I would like to thank my dear mother and brothers for all kinds of supports which give me energy to go further.

I had many interesting discussions regarding studying and living with people at Stony Brook and elsewhere. I would particularly like to thank Ming Xu for his friendship, as well as all kinds of helping and giving good advice during the first four years of living and studying at Stony Brook. I would like to thank Bin Zhang for his generosity.

I would like to thank Professor Mark de Cataldo, Claude LeBrun, Sorin Popescu, and last but certainly not least Paulo Lima-Filho, for helping in studying and offering support when needed.

Thanks are due to my fellow students in Stony Brook, Xiaojun Chen, Haodong Hu, Li Li, Yusuf Mustopa and Dezhen Xu.

In Tianjin, I would particularly like to thank Weiping Zhang for his instruction, encouragement and financial supports during my years as a student in Nankai Institute of Mathematics. It wouldn't have been possible without his help. I also would like to thank Huitao Feng, Enli Guo, Lijing Wang for their help in those years.

My thanks, my love to all.

To my wife, Lihua and my son, Brooks.

Contents

Introduction

0.1 The general context

The purpose of my research is to understand the structure of algebraic cycles on projective varieties by homotopy-theoretic methods, in particular, to develop Lawson homology theory and apply it to tackle problems in algebraic geometry. The homotopy-theoretic approach to algebraic cycles is based on the "Algebraic Suspension Theorem" [L1] and has been developed by E. Friedlander, B. Lawson, P. Mazur, O. Gabber, P. Lima-Filho, and others ([F1], [F2], [FL1], [FL2], [FM], [FG], [L1], [L2], [Li1], etc.). The Lawson homology groups for a complex projective variety are defined by taking homotopy groups of the space of algebraic cycles of a given dimension. They are functorial and yield invariants of varieties up to isomorphism [F1]. They generalize the group of algebraic cycles modulo algebraic equivalence just as the Bloch's higher Chow groups generalize algebraic cycles modulo rational equivalence.

We obtain the following results in my thesis from Chapter 1 to Chapter 7:

Ch1. New birational invariants for a projective manifold are defined by using Lawson homology. These invariants can be highly nontrivial even for projective threefolds. Our techniques involve the weak factorization theorem of Wlodarczyk and tools developed by Friedlander, Lawson, Lima-Filho and others. It is hoped that this will lead to an effective criterion for the irrationality of smooth, rationally connected varieties. A blowup formula for Lawson homology is given in a separate section. As an application, we show that for each $n \geq 5$, there is a smooth rational variety X of dimension n such that the Griffiths groups $\mathrm{Griff}_p(X)$ are infinitely generated even modulo torsion for all p with $2 \leq p \leq n - 3$.

Ch2. In this chapter, we prove that the statement: "The (Generalized) Hodge Conjecture holds for codimension-two cycles on a smooth projective variety X" is a birationally invariant statement, that is, if the statement is true for X, it is also true for all smooth varieties X' which are birationally equivalent to X. We also prove the analogous result for 1-cycles. As direct corollaries, the Hodge Conjecture holds for smooth rational projective manifolds with dimension less than or equal to five, and, the Generalized Hodge Conjecture holds for smooth rational projective manifolds with dimension less than or equal to four.

Ch3. In the first part of this chapter, we show that the assertion "$T_pH_k(X,\mathbb{Q}) = G_pH_k(X,\mathbb{Q})$" (which is called the Friedlander-Mazur conjecture) is a birationally invariant statement for smooth projective varieties X when $p = \dim(X) - 2$ and when $p = 1$. We also establish the Friedlander-Mazur conjecture in certain dimensions. More precisely, for a smooth projective variety X, we show that the topological filtration $T_pH_{2p+1}(X,\mathbb{Q})$ coincides with the geometric filtration $G_pH_{2p+1}(X,\mathbb{Q})$ for all p. (Friedlander and Mazur had previously shown that $T_pH_{2p}(X,\mathbb{Q}) = G_pH_{2p}(X,\mathbb{Q})$. As a corollary, we conclude that for a smooth projective threefold X, $T_pH_k(X,\mathbb{Q}) = G_pH_k(X,\mathbb{Q})$ for all $k \geq 2p \geq 0$ except for the case $p = 1, k = 4$. Finally, we show that the topological and geometric filtrations always coincide if Suslin's conjecture holds.

Ch4. Let X be a smooth projective variety of dimension n on which rational and homological equivalence coincide for algebraic p-cycles in the range $0 \leq p \leq s$. We show that the homologically trivial sector of rational Lawson homology $L_pH_k(X,\mathbb{Q})_{hom}$ vanishes for $0 \leq n-p \leq s+2$. This is an analogue of a theorem of C. Peters in "dual dimensions". Together with Peters' theorem we get that the natural transformation $L_pH_k(X,\mathbb{Q}) \to H_k(X,\mathbb{Q})$ is injective for all p and k when X is a smooth projective variety of dimension 4 and $\mathrm{Ch}_0(X) = \mathbb{Z}$.

Ch5. In this chapter, we construct rational projective 4-dimensional varieties with the property that certain Lawson homology groups tensored with $\mathbb{Q}$ are infinite dimensional $\mathbb{Q}$-vector spaces. More generally, each pair of integers p and k, with $k \geq 0$, $p > 0$, we find a projective variety Y, such that $L_pH_{2p+k}(Y)$ is infinitely generated. This is totally different from the smooth case (cf. [Pe]).

We also construct two singular rational projective 3-dimensional varieties Y and Y' with the same homeomorphism type but different Lawson homology groups, specifically $L_1H_3(Y)$ is not isomorphic to $L_1H_3(Y')$ even up to torsion.

Ch6. In this chapter, we extend Griffiths' Abel-Jacobi map to the homologically trivial part of Lawson homology and showed that this generalization is not trivial in general. As an application, we show that for certain smooth varieties X, the Lawson homology $L_pH_{2p+j}(X)$ is infinitely generated for $j > 0$. In fact for any $p > 0$ and $j > 0$, there exists a smooth variety X such that $L_pH_{2p+j}(X)$ is infinitely generated.

Ch7. By using the notions of "spark" and "differential character" systematically studied by Harvey, Lawson and Zweck ([HLZ]) and the theory of D-bar sparks developed by Harvey and Lawson ([HL2]), we define a homomorphism from Lawson homology to Deligne cohomology for a smooth projective variety. The generalized Abel-Jacobi map we defined in Chapter six is the restriction of this map to the homologically trivial part of the Lawson homology.

The exact statements of the main results and applications will be given below.

0.2 Lawson homology

In this section I briefly describe the basic objects studied in my thesis. Let X be an n-dimensional projective variety defined over $\mathbb{C}$. Let $\mathcal{Z}_p(X)$ be the group of all algebraic p-cycles on X.

The **Lawson homology** $L_pH_k(X)$ of p-cycles is defined by

$$L_pH_k(X) := \pi_{k-2p}(\mathcal{Z}_p(X)) \quad \text{for} \quad k \geq 2p \geq 0,$$

where $\mathcal{Z}_p(X)$ is provided with a natural, compactly generated topology (cf. [F1], [L1]). For convenience, we set (cf. [FHW])

$$L_pH_k(X) := L_0H_k(X), \quad \text{if} \quad p < 0.$$

For general background, the reader is referred to [L2].

In [FM], Friedlander and Mazur showed that there are natural maps, called **cycle class maps**

$$\Phi_{p,k} : L_pH_k(X) \to H_k(X).$$

To state the main results, we need the following definitions:

Definition 0.2.1. *Set*

$$L_pH_k(X)_{hom} := \ker\{\Phi_{p,k} : L_pH_k(X) \to H_k(X)\};$$

$$T_pH_k(X) := \text{Image}\{\Phi_{p,k} : L_pH_k(X) \to H_k(X)\};$$

$$T_pH_k(X, \mathbb{Q}) := T_pH_k(X) \otimes \mathbb{Q}.$$

It was shown in [[FM], §7] that the subspaces $T_pH_k(X, \mathbb{Q})$ form a decreasing filtration:

$$\cdots \subseteq T_pH_k(X, \mathbb{Q}) \subseteq T_{p-1}H_k(X, \mathbb{Q}) \subseteq \cdots \subseteq T_0H_k(X, \mathbb{Q}) = H_k(X, \mathbb{Q})$$

and $T_pH_k(X, \mathbb{Q})$ vanishes if $2p > k$. This is called the **topological filtration**.

Definition 0.2.2. *([[FM],§7]) Denote by*

$$\tilde{F}_pH_k(X, \mathbb{Q}) \subseteq H_k(X, \mathbb{Q})$$

3

the maximal sub-Mixed Hodge structure of span $k - 2p$.(See [Gro] and [FM].) The sub-$\mathbb{Q}$ vector spaces $\tilde{F}_p H_k(X, \mathbb{Q})$ form a decreasing filtration of sub-Hodge structures:

$$\cdots \subseteq \tilde{F}_p H_k(X, \mathbb{Q}) \subseteq \tilde{F}_{p-1} H_k(X, \mathbb{Q}) \subseteq \cdots \subseteq \tilde{F}_0 H_k(X, \mathbb{Q}) \subseteq H_k(X, \mathbb{Q})$$

and $\tilde{F}_p H_k(X, \mathbb{Q})$ vanishes if $2p > k$. This is the homological version of the Hodge filtration.

Definition 0.2.3. *([[FM], §7]) Denote by*

$$G_p H_k(X, \mathbb{Q}) \subseteq H_k(X, \mathbb{Q})$$

the $\mathbb{Q}$-vector subspace of $H_k(X, \mathbb{Q})$ generated by the images of mappings $H_k(Y, \mathbb{Q}) \to H_k(X, \mathbb{Q})$, induced from all morphisms $Y \to X$ of varieties of dimension $\leq k - p$.

The subspaces $G_p H_k(X, \mathbb{Q})$ also form a decreasing filtration:

$$\cdots \subseteq G_p H_k(X, \mathbb{Q}) \subseteq G_{p-1} H_k(X, \mathbb{Q}) \subseteq \cdots \subseteq G_0 H_k(X, \mathbb{Q}) \subseteq H_k(X, \mathbb{Q})$$

called the **geometric filtration**.

It was proved in [FM] that, for any smooth variety X, the topological filtration is finer than the geometric filtration, i.e., $T_p H_k(X, \mathbb{Q}) \subseteq G_p H_k(X, \mathbb{Q})$, for all p and k.

The **Friedlander-Mazur Conjecture**: Let p, k be non-negative integers. For any smooth projective variety X,

$$T_p H_k(X, \mathbb{Q}) = G_p H_k(X, \mathbb{Q}).$$

It was proved in [Gro] that, for any smooth variety X, the geometric filtration is finer than the Hodge filtration, i.e., $G_p H_k(X, \mathbb{Q}) \subseteq \tilde{F}_p H_k(X, \mathbb{Q})$, for all p and k.

The **Hodge Conjecture** (for codimension-q cycles): *The rational cycle class map*

$$cl_q \otimes \mathbb{Q} : \mathcal{Z}^q(X) \otimes \mathbb{Q} \to H^{q,q}(X) \cap H^{2q}(X, \mathbb{Q})$$

is surjective.

The **Hodge Conjecture over $\mathbb{Z}$**: *The rational cycle class map*

$$cl_q : \mathcal{Z}^q(X) \to H^{q,q}(X) \cap \rho(H^{2q}(X, \mathbb{Z}))$$

is surjective.

The **Generalized Hodge Conjecture**: For any smooth projective variety X,

$$G_p H_k(X, \mathbb{Q}) = \tilde{F}_p H_k(X, \mathbb{Q})$$

4

for all p and k. Using the notation given in [Lew1], we denote by $\widetilde{GHC}(p, k, X)$ the assertion that this conjecture holds for given p, k and X.

For convenience later, we recall the following definition:

Definition 0.2.4. *A smooth projective variety X over $\mathbb{C}$ is called* **rationally connected** *if there is a rational curve through any 2 points of X. A necessary condition for Z to be rationally connected is that* $\mathrm{Ch}_0(X) \cong \mathbb{Z}$.

For equivalent descriptions of this definition, see the paper of Kollár, Miyaoka and Mori [KMM].

0.3 The main results

In this section I shall give a detailed presentation of the results summarized in section 0.1.

Theorem 0.3.1. *If X is a smooth n-dimensional projective variety, then $L_1 H_k(X)_{hom}$ and $L_{n-2}H_k(X)_{hom}$ are smooth birational invariants for X. More precisely, if $\varphi : X \to X'$ is a birational map between smooth projective manifolds X and X', then φ induces isomorphisms $L_1 H_k(X)_{hom} \cong L_1 H_k(X')_{hom}$ for $k \geq 2$ and $L_{n-2}H_k(X)_{hom} \cong L_{n-2}H_k(X')_{hom}$ for $k \geq 2(n-2)$. In particular, $L_1 H_k(X)_{hom} = 0$ and $L_{n-2}H_k(X)_{hom} = 0$ for any smooth rational variety X.*

Corollary 0.3.1. *Let X be a smooth rational projective variety with $\dim(X) \leq 4$, then $\Phi_{p,k} : L_p H_k(X) \to H_k(X)$ is injective for all $k \geq 2p \geq 0$.*

Remark 0.3.1. *In general, for $2 \leq p \leq n-3$, $L_p H_k(X)_{hom}$ is* **not** *a birational invariant for the smooth projective variety X. This follows from the blowup formula in Lawson homology given in Theorem 0.3.2 below.*

Remark 0.3.2. *If $p = 0, n-1, n$, then $L_p H_k(X)_{hom} = 0$ for all $k \geq 2p$. In these cases, the statement in the theorem is trivial. The case for $p = 0$ follows from the Dold-Thom theorem. The case for $p = n-1$ is due to Friedlander [F1]. The case for $p = n$ follows from the definition.*

Using the notation in section 0.2, we have the following:

Theorem 0.3.2. *(Lawson homology for a blowup) Let X be smooth projective manifold and $Y \subseteq X$ a smooth subvariety of codimension r. Let $\sigma : \tilde{X}_Y \to X$ be the blowup of X along Y, $\pi : D = \sigma^{-1}(Y) \to Y$ the natural map, and $\imath : D = \sigma^{-1}(Y) \to \tilde{X}_Y$ the exceptional divisor. Then for each p, we have*

$$L_p H_k(\tilde{X}_Y) \cong \left\{ \bigoplus_{1 \leq j \leq r-1} L_{p-j} H_{k-2j}(Y) \right\} \oplus L_p H_k(X)$$

5

From the blowup formula for Lawson homology and Clemens' result [Cl], we have the following

Corollary 0.3.2. *For each $n \geq 5$, there exists rational manifold X with $\dim(X) = n$ such that*

$$\dim_{\mathbb{Q}} \left\{ \mathrm{Griff}_p(X) \otimes \mathbb{Q} \right\} = \infty, \quad 2 \leq p \leq n - 3.$$

We can apply our method to obtain the following:

Theorem 0.3.3. *Let X be a smooth projective variety. If the Hodge conjecture for codimension 2 cycles over $\mathbb{Z}$ holds for X, i.e., if we have $\mathrm{Hodge}^{2,2}(X, \mathbb{Z})$, then it holds for any smooth projective variety X' birational to X. That is, $\mathrm{Hodge}^{2,2}(X, \mathbb{Z})$ is a birationally invariant assertion for smooth varieties X.*

As a corollary, we have

Corollary 0.3.3. *If X is a rational manifold with $\dim(X) \leq 5$, then the Hodge conjecture $\mathrm{Hodge}^{p,p}(X, \mathbb{Q})$ is true for $1 \leq p \leq \dim(X)$. In fact, $\mathrm{Hodge}^{p,p}(X', \mathbf{Z})$ is true except possibly for $p = 3, \dim(X) = 5$.*

Remark 0.3.3. *By using the technique of the diagonal decomposition, Bloch and Srinivas [BS] showed that $\mathrm{Hodge}^{2,2}(X, \mathbb{Q})$ holds if the Chow group of 0-cycles $\mathrm{Ch}_0(X) \cong \mathbb{Z}$ for any smooth projective variety X. Laterveer [Lat] generalized this technique and showed the Hodge conjecture holds for a class of projective manifolds with small Chow groups.*

More generally, we have

Theorem 0.3.4. *"$\widetilde{GHC}(n - 2, k, X)$" is a birationally invariant property of smooth n-dimensional varieties X when $k \geq 2(n-2)+1$. More precisely, if $\widetilde{GHC}(n-2, k, X)$ holds for a smooth variety X, then for any smooth variety X' birational to X, $\widetilde{GHC}(n-2, k, X')$ holds.*

Similarly we can show that

Theorem 0.3.5. *$\widetilde{GHC}(1, k, X)$ for X for any integer $k \geq 2$ is a birationally invariant property of smooth varieties X.*

Corollary 0.3.4. *For any smooth rational variety X with $\dim(X) \leq 4$, the Generalized Hodge Conjecture holds.*

Applying the method to the topological filtration, we obtain the following:

Theorem 0.3.6. *Let X be a smooth projective variety of dimension n. If*

$$T_p H_k(X, \mathbb{Q}) = G_p H_k(X, \mathbb{Q})$$

for $p = 1$, (resp. $p = n - 2$) and $k \geq 2p$, then this also holds for any smooth projective variety X' which is birationally equivalent to X with $p = 1$, (resp. $p = n - 2$) and $k \geq 2p$.

The following theorem takes a result of Friedlander and Mazur [FM] one step further:

Theorem 0.3.7. The **Friedlander-Mazur Conjecture** holds for $k = 2p + 1$. That is, for any smooth projective variety X,

$$T_p H_{2p+1}(X, \mathbb{Q}) = G_p H_{2p+1}(X, \mathbb{Q}).$$

As corollaries, we have shown the **Friedlander-Mazur Conjecture** in the following cases:

Corollary 0.3.5. Let X be a smooth projective threefold. Then $T_p H_k(X, \mathbb{Q}) = G_p H_k(X, \mathbb{Q})$ for all $k \geq 2p \geq 0$ except for the case $p = 1, k = 4$.

Corollary 0.3.6. Let X be a smooth projective threefold with $H^{2,0}(X) = 0$. Then $T_p H_k(X, \mathbb{Q}) = G_p H_k(X, \mathbb{Q})$ for any $k \geq 2p \geq 0$. This holds whenever X is a smooth complete intersection of dimension 3.

By using the Künneth formula in homology, we have

Corollary 0.3.7. Let X be the product of a smooth projective curve and a smooth simply connected projective surface. Then $T_p H_k(X, \mathbb{Q}) = G_p H_k(X, \mathbb{Q})$ for any $k \geq 2p \geq 0$.

Corollary 0.3.8. For a smooth projective fourfold X, the assertion that

$$T_p H_k(X, \mathbb{Q}) = G_p H_k(X, \mathbb{Q})$$

holds for all $k \geq 2p \geq 0$ is a birationally invariant statement. In particular, if X is a rational manifold with $\dim(X) \leq 4$, then this assertion holds for any $k \geq 2p \geq 0$.

Remark 0.3.4. A Conjecture given by Suslin (cf. [FHW], §7) implies that $L_p H_{n+p}(X^n) \cong H_{n+p}(X^n)$ for any $p \geq 0$.

As an application of Theorem 0.3.7, we have the following result:

Corollary 0.3.9. If the Suslin's Conjecture is true, then the topological filtration is the same as the geometric filtration for any smooth projective variety.

We generalize the higher Abel-Jacobi map introduced by Griffiths to Lawson homology and obtain the following result:

Theorem 0.3.8. Let X be a smooth projective variety. There is a well-defined map

$$\Phi : L_p H_{2p+k}(X)_{hom} \rightarrow \left\{ \bigoplus_{r > k+1, r+s=k+1} H^{p+r,p+s}(X) \right\}^* \Big/ H_{2p+k+1}(X, \mathbb{Z})$$

which generalizes the Griffiths' higher Abel-Jacobi map defined in [G]. Moreover, for any $p > 0$ and $k \geq 0$, we can find examples of smooth projective varieties such that the image of this map is infinitely generated.

Theorem 0.3.9. For any $k \geq 0$, there exists a projective manifold X of dimension $k + 3$ such that $L_1 H_{k+2}(X)_{hom} \otimes \mathbb{Q}$ is nontrivial, in fact, infinite dimensional over $\mathbb{Q}$.

By applying the projective bundle theorem in [FG], we have the following result:

Corollary 0.3.10. For any $p > 0$ and $k \geq 0$, there exists a projective manifold X such that $L_p H_{k+2p}(X)_{hom} \otimes \mathbb{Q}$ is an infinite dimensional vector space over $\mathbb{Q}$.

Theorem 0.3.10. Let X be a smooth projective manifold with dimension n. There exists a well-defined homomorphism to Deligne cohomology

$$\hat{a} : L_p H_{k+2p}(X) \to H_{\mathcal{D}}^{2(n-p)-k}(X, \mathbb{Z}(n - p - k - 1)),$$

whose restriction to $L_p H_k(X)_{hom}$ coincides with the generalized Abel-Jacobi map defined in Theorem 0.3.8 above. Furthermore, the cycle class map $\Phi_{p,k}$ factors through $\hat{a}$.

Using results in Peters' paper [Pe], we have the following:

Theorem 0.3.11. Let X be a smooth projective variety of dimension n for which rational and homological equivalence coincide for $p-$cycles in the range $0 \leq p \leq s$. Then $L_p H_*(X)_{hom} \otimes \mathbb{Q} = 0$ in the range $0 \leq n - p \leq s + 2$.

Corollary 0.3.11. Let X be a smooth projective variety with $\dim(X) = 4$ and $Ch_0(X) \cong Z$. Then $L_p H_k(X)_{hom} \otimes \mathbb{Q} = 0$ for all p and k. In particular, all the smooth hypersurfaces of dimension 4 with degree less than or equal to 5 have this property (cf. [Ro]).

Remark 0.3.5. Lawson homology could give a criterion for the difference between **rationality** and **rational connectivity**. For example, Corollary 3.1 tells us that for any smooth projective rational variety X of $\dim(X) = 4$, $L_p H_k(X)_{hom} = 0$ for any p and k. Hence the nontriviality of any $L_p H_k(X)_{hom}$ for a rationally connected fourfold X would imply the irrationality of X.

Based on a result of Clemens [Cl], we get the following:

Theorem 0.3.12. There exists a **rational** projective 4-dimensional variety X such that $L_1 H_3(X) \otimes \mathbb{Q}$ is **not** a finite dimensional $\mathbb{Q}$-vector space.

Chapter 1

Birational invariants defined by Lawson homology

1.1 Introduction

In this chapter, all varieties are defined over $\mathbb{C}$. Let X be an n-dimensional projective variety. The **Lawson homology** $L_p H_k(X)$ of p-cycles is defined by

$$L_p H_k(X) := \pi_{k-2p}(\mathcal{Z}_p(X)) \quad for \quad k \geq 2p \geq 0,$$

where $\mathcal{Z}_p(X)$ is provided with a natural topology (cf. [F1], [L1]). For general background, the reader is referred to [L2].

In [FM], Friedlander and Mazur showed that there are natural transformations, called **cycle class maps**

$$\Phi_{p,k} : L_p H_k(X) \to H_k(X).$$

Define

$$L_p H_k(X)_{hom} := \ker\{\Phi_{p,k} : L_p H_k(X) \to H_k(X)\}.$$

The **Griffiths group** of codimension q-cycles is defined to

$$\mathrm{Griff}^q(X) := \mathcal{Z}^q(X)_{hom}/\mathcal{Z}^q(X)_{alg}$$

It was proved by Friedlander [F1] that, for any smooth projective variety X,

$$L_p H_{2p}(X) \cong \mathcal{Z}_p(X)/\mathcal{Z}_p(X)_{alg}.$$

9

Therefore
$$L_p H_{2p}(X)_{hom} \cong \mathrm{Griff}_p(X),$$
where $\mathrm{Griff}_p(X) := \mathrm{Griff}^{n-p}(X)$.

The main result in this chapter is the following

Theorem 1.1.1. *If X is a smooth n-dimensional projective variety, then $L_1 H_k(X)_{hom}$ and $L_{n-2} H_k(X)_{hom}$ are smooth birational invariants for X. More precisely, if $\varphi : X \to X'$ is a birational map between smooth projective manifolds X and X', then φ induces isomorphisms $L_1 H_k(X)_{hom} \cong L_1 H_k(X')_{hom}$ for $k \geq 2$ and $L_{n-2} H_k(X)_{hom} \cong L_{n-2} H_k(X')_{hom}$ for $k \geq 2(n-2)$. In particular, $L_1 H_k(X)_{hom} = 0$ and $L_{n-2} H_k(X)_{hom} = 0$ for any smooth rational variety.*

Corollary 1.1.1. *Let X be a smooth rational projective variety with $\dim(X) \leq 4$, then $\Phi_{p,k} : L_p H_k(X) \to H_k(X)$ is injective for all $k \geq 2p \geq 0$.*

Remark 1.1.1. *In general, for $2 \leq p \leq n-3$, $L_p H_k(X)_{hom}$ is not a birational invariant for the smooth projective variety X. This follows from the blowup formula in Lawson homology (See Corollary 1.1.2, 1.1.3).*

Remark 1.1.2. *If $p = 0, n-1, n$, then $L_p H_k(X)_{hom} = 0$ for all $k \geq 2p$. In these cases, the statement in Theorem 1.1.1 is trivial. The case for $p = 0$ follows from Dold-Thom theorem ([DT]). The case for $p = n-1$ is due to Friedlander [F1]. The case for $p = n$ is from the definition. In particular, these invariants are trivial for smooth projective varieties with dimension less than or equal to two.*

Another result is this chapter is the following:

Theorem 1.1.2. *(Lawson homology for a blowup) Let X be smooth projective manifold and $Y \subset X$ be a smooth subvariety of codimension r. Let $\sigma : \tilde{X}_Y \to X$ be the blowup of X along Y, $\pi : D = \sigma^{-1}(Y) \to Y$ the natural map, and $i : D = \sigma^{-1}(Y) \to \tilde{X}_Y$ the exceptional divisor of the blowing up. Then for each p, k with $k \geq 2p \geq 0$, we have the following isomorphism*

$$I_{p,k} : \left\{ \bigoplus_{1 \leq j \leq r-1} L_{p-j} H_{k-2j}(Y) \right\} \bigoplus L_p H_k(X) \cong L_p H_k(\tilde{X}_Y)$$

As applications, we have the following

10

Corollary 1.1.2. *For each $n \geq 5$, there exists a* rational *manifold X with $\dim(X) = n$ such that*

$$\dim_{\mathbb{Q}}\{\mathrm{Griff}_p(X) \otimes \mathbb{Q}\} = \infty, \quad 2 \leq p \leq n - 3.$$

Corollary 1.1.3. *For any integer $p > 1$ and $k \geq 0$, there exists* rational *projective manifold X such that $L_p H_{k+2p}(X) \otimes \mathbb{Q}$ is an infinite dimensional vector space over $\mathbb{Q}$.*

The main tools used to prove the main result are: the long exact localization sequence given by Lima-Filho in [Li1], the explicit formula for the Lawson homology of codimension-one cycles on a smooth projective manifold given by Friedlander in [F1], and the weak factorization theorem proved by Wlodarczyk and others in [Wl] and in [AKMW].

1.2 Some fundamental materials in Lawson homology

First recall that for a morphism $f : U \to V$ between projective varieties, there exist induced homomorphism

$$f_* : L_p H_k(U) \to L_p H_k(V)$$

for all $k \geq 2p \geq 0$, and if $g : V \to W$ is another morphism between projective varieties, then

$$(g \circ f)_* = g_* \circ f_*.$$

Furthermore, it has been shown by C. Peters [Pe] that if U and V are smooth and projective, there are Gysin "wrong way" homomorphisms $f^* : L_p H_k(V) \to L_{p-c} H_{k-2c}(U)$, where $c = \dim(V) - \dim(U)$. If $g : V \to W$ is another morphism between smooth projective varieties, then

$$(g \circ f)^* = f^* \circ g^*.$$

Recall also the fact that there is a long exact sequence (cf. [Li1], also [FG])

$$\cdots \to L_p H_k(U - V) \to L_p H_k(U) \to L_p H_k(V) \to L_p H_{k-1}(U - V) \to \cdots,$$

where U is quasi-projective and $U - V$ is any algebraic closed subset in U.

Let X be a smooth projective variety and $i_0 : Y \hookrightarrow X$ a smooth subvariety of codimension $r \geq 2$. Let $\sigma : \tilde{X}_Y \to X$ be the blowup of X along Y, $\pi : D = \sigma^{-1}(Y) \to Y$ the natural map, and $i : D = \sigma^{-1}(Y) \hookrightarrow \tilde{X}_Y$ the exceptional divisor of the blowup. Set $U := X - Y \cong \tilde{X}_Y - D$. Denote by j_0 the inclusion $U \subset X$ and j the inclusion $U \subset \tilde{X}_Y$. Note that $\pi : D = \sigma^{-1}(Y) \to Y$ makes D into a projective bundle of rank $r - 1$, given precisely by $D = \mathrm{P}(N_{Y/X})$ and we have (cf. [[V2], pg. 271])

$$\mathcal{O}_{\tilde{X}_Y}(D)|_D = \mathcal{O}_{\mathrm{P}(N_{Y/X})}(-1).$$

Denote by h the class of $\mathcal{O}_{\mathrm{P}(N_{Y/X})}(-1)$ in $\mathrm{Pic}(D)$. We have $h = -D|_D$ and $-h = i^* i_* : L_q H_m(D) \to L_{q-1} H_{m-2}(D)$ for $0 \leq 2q \leq m$ ([FG], Theorem 2.4], [[Pe], Lemma 11]). The last equality can be equivalently regarded as a Lefschetz operator

$$- h = i^* i_* : L_q H_m(D) \to L_{q-1} H_{m-2}(D), \quad 0 \leq 2q \leq m. \tag{1.1}$$

The proof of the main result is based on the following lemmas:

Lemma 1.2.1. *For each $p \geq 0$, we have the following commutative diagram*

$$
\begin{array}{ccccccccc}
\cdots \to & L_p H_k(D) & \xrightarrow{i_*} & L_p H_k(\tilde{X}_Y) & \xrightarrow{j^*} & L_p H_k(U) & \xrightarrow{\delta} & L_p H_{k-1}(D) & \to & \cdots \\
& \downarrow \pi_* & & \downarrow \sigma_* & & \downarrow \cong & & \downarrow \pi_* & & \\
\cdots \to & L_p H_k(Y) & \xrightarrow{(i_0)^*} & L_p H_k(X) & \xrightarrow{j_0^*} & L_p H_k(U) & \xrightarrow{(\delta_0)^*} & L_p H_{k-1}(Y) & \to & \cdots
\end{array}
$$

Proof. This is from the corresponding commutative diagram of fibration sequences of p-cycles. More precisely, to show the first square, we begin from the following commutative diagram

$$
\begin{array}{ccc}
D & \xrightarrow{i} & \tilde{X}_Y \\
\downarrow \pi & & \downarrow \sigma \\
Y & \xrightarrow{i_0} & X.
\end{array}
$$

From this, we obtain the corresponding commutative diagram of p-cycles:

$$
\begin{array}{ccc}
\mathcal{Z}_p(D) & \xrightarrow{i_*} & \mathcal{Z}_p(\tilde{X}_Y) \\
\downarrow \pi_* & & \downarrow \sigma_* \\
\mathcal{Z}_p(Y) & \xrightarrow{(i_0)_*} & \mathcal{Z}_p(X).
\end{array}
$$

12

Since Y is a smooth projective variety, $\tilde{X}_Y$ and D are smooth projective varieties, we have the following commutative diagram

$$
\begin{array}{ccc}
\mathcal{Z}_p(\tilde{X}_Y) & \rightarrow & \mathcal{Z}_p(\tilde{X}_Y)/\mathcal{Z}_p(D) \\
\downarrow \sigma_* & & \downarrow \cong \\
\mathcal{Z}_p(X) & \rightarrow & \mathcal{Z}_p(X)/\mathcal{Z}_p(Y).
\end{array}
$$

Therefore we obtain the following commutative diagram of the fibration sequences of p-cycles

$$
\begin{array}{ccccc}
\mathcal{Z}_p(D) & \overset{i_*}{\hookrightarrow} & \mathcal{Z}_p(\tilde{X}_Y) & \rightarrow & \mathcal{Z}_p(\tilde{X}_Y)/\mathcal{Z}_p(D) \\
\downarrow \pi_* & & \downarrow \sigma_* & & \downarrow \cong \\
\mathcal{Z}_p(Y) & \overset{(i_0)_*}{\hookrightarrow} & \mathcal{Z}_p(X) & \rightarrow & \mathcal{Z}_p(X)/\mathcal{Z}_p(Y).
\end{array}
$$

where the fact that the rows are fibration sequences is due to Lima- Filho [Li1].

By taking the homotopy groups of these fibration sequences, we get the long exact sequences of commutative diagram given in the Lemma.

$\square$

Proposition 1.2.1. *If $p = 0$, then we have the following commutative diagram*

$$
\begin{array}{ccccccccc}
\cdots \rightarrow & H_k(D) & \overset{i_*}{\rightarrow} & H_k(\tilde{X}_Y) & \overset{j^*}{\rightarrow} & H_k^{BM}(U) & \overset{\delta_*}{\rightarrow} & H_{k-1}(D) & \rightarrow & \cdots \\
& \downarrow \pi_* & & \downarrow \sigma_* & & \downarrow \cong & & \downarrow \pi_* & & \\
\cdots \rightarrow & H_k(Y) & \overset{(i_0)^*}{\rightarrow} & H_k(X) & \overset{j_0^*}{\rightarrow} & H_k^{BM}(U) & \overset{(\delta_0)_*}{\rightarrow} & H_{k-1}(Y) & \rightarrow & \cdots
\end{array}
$$

Moreover, if $x \in H_k(D)$ maps to zero under π_* and i_*, then $x = 0 \in H_k(D)$.

Proof. The first conclusion follows directly from Lemma 1.2.1 with $p = 0$ and the Dold-Thom Theorem. For the second conclusion assume $i_*(x) = 0$ and $\pi_*(x) = 0$. Then there exists an element $y \in H_{k+1}^{BM}(U)$ such that the image of y under the boundary map $(\delta_0)_* : H_{k+1}^{BM}(U) \rightarrow H_k(Y)$ is 0 by the given condition. Hence there exists an element $z \in H_{k+1}(X)$ such that $(j_0)^*(z) = y$. Now the surjectivity of the map $\sigma_* : H_{k+1}(\tilde{X}_Y) \rightarrow H_{k+1}(X)$ implies that there is an element $\tilde{z} \in H_{k+1}(\tilde{X}_Y)$ such that $j^*(\tilde{z}) = y$. Therefore, $x = 0 \in H_k(D)$.

$\square$

Corollary 1.2.1. *If $p = n - 2$, then we have the following commutative diagram*

$$\cdots \to \quad L_{n-2}H_k(D) \quad \overset{i_*}{\to} \quad L_{n-2}H_k(\tilde{X}_Y) \quad \overset{j^*}{\to} \quad L_{n-2}H_k(U) \quad \overset{\delta_*}{\to} \quad L_{n-2}H_{k-1}(D) \quad \to \quad \cdots$$
$$\downarrow \pi_* \qquad\qquad \downarrow \sigma_* \qquad\qquad \downarrow\cong \qquad\qquad \downarrow \pi_*$$
$$\cdots \to \quad L_{n-2}H_k(Y) \quad \overset{(i_0)_*}{\to} \quad L_{n-2}H_k(X) \quad \overset{j_0^*}{\to} \quad L_{n-2}H_k(U) \quad \overset{(\delta_0)_*}{\to} \quad L_{n-2}H_{k-1}(Y) \quad \to \quad \cdots$$

Lemma 1.2.2. *For each p, we have the following commutative diagram*

$$\cdots \to \quad L_pH_k(D) \quad \overset{i_*}{\to} \quad L_pH_k(\tilde{X}_Y) \quad \overset{j^*}{\to} \quad L_pH_k(U) \quad \overset{\delta_*}{\to} \quad L_pH_{k-1}(D) \quad \to \quad \cdots$$
$$\downarrow \Phi_{p,k} \qquad\quad \downarrow \Phi_{p,k} \qquad\quad \downarrow \Phi_{p,k} \qquad\quad \downarrow \Phi_{p,k-1}$$
$$\cdots \to \quad H_k(D) \quad \overset{i_*}{\to} \quad H_k(\tilde{X}_Y) \quad \overset{j^*}{\to} \quad H_k^{BM}(U) \quad \overset{\delta_*}{\to} \quad H_{k-1}(D) \quad \to \quad \cdots$$

In particular, it is true for $p = 1, n - 2$.

Proof. See [Li1] and also [FM].

$\square$

Lemma 1.2.3. *For each p, we have the following commutative diagram*

$$\cdots \to \quad L_pH_k(Y) \quad \overset{(i_0)_*}{\to} \quad L_pH_k(X) \quad \overset{j^*}{\to} \quad L_pH_k(U) \quad \overset{(\delta_0)_*}{\to} \quad L_pH_{k-1}(Y) \quad \to \quad \cdots$$
$$\downarrow \Phi_{p,k} \qquad\quad \downarrow \Phi_{p,k} \qquad\quad \downarrow \Phi_{p,k} \qquad\quad \downarrow \Phi_{p,k-1}$$
$$\cdots \to \quad H_k(Y) \quad \overset{(i_0)_*}{\to} \quad H_k(X) \quad \overset{j^*}{\to} \quad H_k^{BM}(U) \quad \overset{(\delta_0)_*}{\to} \quad H_{k-1}(Y) \quad \to \quad \cdots$$

In particular, it is true for $p = 1, n - 2$.

Proof. See [Li1] and also [FM].

$\square$

1.3 Lawson homology for blowups

As an application of Lemma 1.2.1, we give an explicit formula for a blowup in Lawson homology. Since it may have some independent interest, we devote a separate section to it. First, we want to revise the projective bundle theorem given by Friedlander and Gabber ([FG], Prop.2.5). It is convenient to extend the definition of Lawson homology by setting

$$L_pH_k(X) = L_0H_k(X), \quad if \quad p < 0.$$

Now we have the following revised "Projective Bundle Theorem":

Proposition 1.3.1. *Let E be an algebraic vector bundle of rank r over a smooth projective variety Y, then for each $p \geq 0$ we have*

$$L_p H_k(\mathrm{P}(E)) \cong \bigoplus_{j=0}^{r-1} L_{p-j} H_{k-2j}(Y)$$

where $\mathrm{P}(E)$ is the projectivization of the vector bundle E.

Remark 1.3.1. *The difference between this and the projective bundle theorem of [FG] is that here we place no restriction on p.*

Proof. For $p \geq r - 1$, this is exactly the projective bundle theorem given in [FG]. If $p < r - 1$, we have the same method of [FG], i.e., the localization sequence and the naturality of Φ, to reduce to the case in which E is trivial. From

$$\mathcal{Z}_0(\mathrm{P}^{r-1} \times Y) \to \mathcal{Z}_0(\mathrm{P}^r \times Y) \to \mathcal{Z}_0(\mathbb{C}^r \times Y),$$

we have the long exact localization sequence given at the beginning of section 2:

$$\cdots \to L_0 H_k(\mathrm{P}^{r-1} \times Y) \to L_0 H_k(\mathrm{P}^r \times Y) \to L_0 H_k(\mathbb{C}^r \times Y) \to L_0 H_{k-1}(\mathrm{P}^{r-1} \times Y) \to \cdots .$$

From this, and the Künneth formula for $\mathrm{P}^r \times Y$, we have the following isomorphism:

$$(*) \quad H_{k-2r}(Y) \cong L_0 H_k(\mathbb{C}^r \times Y) \cong H_k^{BM}(\mathbb{C}^r \times Y).$$

Note that

$$(**) \quad H_{k-2r}(Y) \cong L_{p-r} H_{k-2r}(Y) \quad \text{if} \quad p \leq r.$$

All the remaining arguments are the same as those in [[FG], Prop. 2.5], as we review in the following.

We want to use induction on r. For $r - 1 = p$, the conclusion holds. From the commutative diagram of abelian groups of cycles:

$$\{\oplus_{j=0}^{p} \mathcal{Z}_{p-j}(X)\} \bigoplus \{\oplus_{j=p+1}^{r-1} \mathcal{Z}_0(X \times \mathbb{C}^{j-p})\} \quad \to \quad \{\oplus_{j=0}^{p} \mathcal{Z}_{p-j}(X)\} \bigoplus \{\oplus_{j=p+1}^{r} \mathcal{Z}_0(X \times \mathbb{C}^{j-p})\}$$
$$\downarrow \qquad\qquad\qquad\qquad\qquad\qquad\qquad \downarrow$$
$$\mathcal{Z}_p(X \times \mathrm{P}^{r-1}) \qquad\qquad\qquad \to \qquad\qquad \mathcal{Z}_p(X \times \mathrm{P}^r)$$

We obtain the commutative diagram of fibration sequences:

15

$$\{\oplus_{j=0}^p \mathcal{Z}_{p-j}(X)\} \bigoplus \{\oplus_{j=p+1}^{r-1} \mathcal{Z}_{p-j}(X)\} \quad \rightarrow \quad \{\oplus_{j=0}^p \mathcal{Z}_{p-j}(X)\} \bigoplus \{\oplus_{j=p+1}^r \mathcal{Z}_{p-j}(X)\}$$

$$\downarrow \qquad\qquad\qquad\qquad\qquad\qquad\qquad \downarrow$$

$$\mathcal{Z}_p(X \times \mathrm{P}^{r-1}) \qquad\qquad \rightarrow \qquad\qquad \mathcal{Z}_p(X \times \mathrm{P}^r)$$

$$\rightarrow \qquad\qquad \mathcal{Z}_0(X \times \mathbb{C}^{r-p})$$

$$\downarrow$$

$$\rightarrow \qquad\qquad \mathcal{Z}_p(X \times \mathbb{C}^r)$$

where $\mathcal{Z}_{p-j}(X) := \mathcal{Z}_0(X \times \mathbb{C}^{j-p})$ for $p - j < 0$.

The first vertical arrow is a homotopy equivalence by induction. The last one is a homotopy equivalence by Complex Suspension Theorem [L1]. Hence by the Five Lemma, we obtain the homotopy equivalence of the middle one.

The proof is completed by combining this with (*) and (**) above.

$\square$

Remark 1.3.2. *The isomorphism*

$$\psi : \bigoplus_{j=0}^{r-1} L_{p-j} H_{k-2j}(Y) \xrightarrow{\cong} L_p H_k(\mathrm{P}(E))$$

in Proposition 1.3.1 is given explicitly by

$$\psi(u_0, u_1, \cdots, u_{r-1}) = \sum_{j=0}^{r-1} h^j \pi^* u_j$$

where h is the Lefschetz hyperplane operator

$$h : L_q H_m(\mathrm{P}(E)) \rightarrow L_{q-1} H_{m-2}(\mathrm{P}(E))$$

defined in 1.1. For $p \geq r-1$, this explicit formula has been proved in [[FG], Prop. 2.5]. In the remaining cases, h is the Lefschetz hyperplane operator $h : H_m(\mathrm{P}(E)) \rightarrow H_{m-2}(\mathrm{P}(E))$ defined in 1.1.

Using the notations in section 2, we have the following:

Theorem 1.3.1. *(Lawson homology for a blowup) Let X be smooth projective manifold* and $Y \subset X$ be a smooth subvariety of codimension r. Let $\sigma : \tilde{X}_Y \to X$ be the blowup of X along Y, $\pi : D = \sigma^{-1}(Y) \to Y$ the natural map, and $i : D = \sigma^{-1}(Y) \to \tilde{X}_Y$ the exceptional divisor of the blowing up. Then for each p, k with $k \geq 2p \geq 0$, we have the following isomorphism

$$I_{p,k} : \left\{ \bigoplus_{1 \leq j \leq r-1} L_p{}_{-j} H_{k-2j}(Y) \right\} \oplus L_p H_k(X) \xrightarrow{\cong} L_p H_k(\tilde{X}_Y)$$

given by

$$I_{p,k}(u_1, \cdots, u_{r-1}, u) = \sum_{j=1}^{r-1} i_* h^j \pi^* u_j + \sigma^* u$$

Proof. We use certain idea of the proof of Chow groups for blowups. Let $U := \tilde{X}_Y - D = X - Y$. By the definitions of the maps i, π and σ, and Lemma 1.2.1, we have the following commutative diagram of the long exact localization sequences:

$$
\begin{array}{ccccccccc}
\cdots \to & L_p H_k(D) & \xrightarrow{i_*} & L_p H_k(\tilde{X}_Y) & \xrightarrow{j^*} & L_p H_k(U) & \xrightarrow{\delta_*} & L_p H_{k-1}(D) & \to & \cdots \\
 & \downarrow \pi_* & & \downarrow \sigma_* & & \downarrow \cong & & \downarrow \pi_* & & \\
\cdots \to & L_p H_k(Y) & \xrightarrow{(i_0)_*} & L_p H_k(X) & \xrightarrow{j_0^*} & L_p H_k(U) & \xrightarrow{(\delta_0)_*} & L_p H_{k-1}(Y) & \to & \cdots
\end{array}
\tag{1.2}
$$

From this, and the surjectivity of j^*, we have

$$L_p H_{2p}(\tilde{X}_Y) = \sigma^* L_p H_{2p}(X) + i_* L_p H_{2p}(D).$$

By the "revised" projective bundle theorem above, for any $p \geq 0$, there is an isomorphism

$$L_p H_k(D) \cong \bigoplus_{j=0}^{r-1} h^j \pi^* L_{p-j} H_{k-2j}(Y), \quad 0 \leq 2p \leq k.$$

Hence we see that

$$L_p H_{2p}(\tilde{X}_Y) = \sigma^* L_p H_{2p}(X) + \Sigma_{j=0}^{r-1} i_* h^j \pi^* L_{p-j} H_{2p-2j}(Y). \tag{1.3}$$

But clearly by Lemma 1.2.1 and the projective bundle theorem, if $u \in L_p H_k(Y)$, then

$$\sigma_*(i_* h^{r-1} \pi^*(u)) = (i_0)_*(u).$$

17

Since σ is a birational morphism, it has degree one. As a directly corollary of the projection formula (cf. [Pe], Lemma 11 c.), we have $\sigma_*(\sigma^*a) = a$ for any $a \in L_pH_k(X)$. We have

$$\sigma_*(\sigma^*((i_0)_*u)) = (i_0)_*u, \quad u \in L_pH_k(Y).$$

Thus we obtain the relations

$$v := i_*h^{r-1}\pi^*u - \sigma^*((i_0)_*u) \in \ker \sigma_*, \quad u \in L_pH_k(Y)$$

Since $j^* = (j_0)^*\sigma_*$ in (1.2), we get $j^*(v) = 0$. From the exactness of the upper row in (1.2), we get

$$v \in \sum_{j=1}^{r-1} i_*h^j L_{p-j}H_{k-2j}(Y). \tag{1.4}$$

The equality (1.3) and the relation (1.4) together imply immediately that the map $I_{p,2p}$ is surjective for the case $k = 2p$.

To prove the injectivity for the case that $k = 2p$, consider

$$(u_1, u_2, \cdots, u_{r-1}, u) \in \ker I_{p,2p}.$$

Applying σ_*, we find that $u = 0$. Note that $i^*i_* = -h$. Now applying i^* to the equality

$$\sum_{j=1}^{r-1} i_*h^j\pi^*u_j = 0,$$

we get

$$\sum_{j=1}^{r-1} h^{j+1}\pi^*u_j = 0 \in L_{p-1}H_{k-2}(D).$$

The isomorphism in Proposition 1.3.1 implies that $u_j = 0$ for $1 \le j \le r - 1$. This completes the proof for the case $k = 2p$.

From this and (1.2), we have

$$
\begin{array}{ccccccccc}
\cdots \to & L_pH_{2p+1}(D) & \xrightarrow{i_*} & L_pH_{2p+1}(\tilde{X}_Y) & \xrightarrow{j^*} & L_pH_{2p+1}(U) & \xrightarrow{\delta_*} & 0 & \\
& \downarrow \pi_* & & \downarrow \sigma_* & & \downarrow \cong & & & (1.5) \\
\cdots \to & L_pH_{2p+1}(Y) & \xrightarrow{(i_0)_*} & L_pH_{2p+1}(X) & \xrightarrow{j_0^*} & L_pH_{2p+1}(U) & \xrightarrow{(\delta_0)_*} & 0 &
\end{array}
$$

Now the situation for $k = 2p + 1$ is the same as that in the case $k = 2p$. From (1.5) and the "revised" projective bundle theorem, we have

$$L_p H_{2p+1}(\tilde{X}_Y) = \sigma^* L_p H_{2p+1}(X) + \Sigma_{j=0}^{r-1} i_* h^j \pi^* L_{p-j} H_{2p+1-2j}(Y). \qquad (1.6)$$

From (1.4) and (1.6), we obtain the surjectivity of $I_{p,2p+1}$ for the case that $k = 2p + 1$.

To prove the injectivity, consider $(u_1, u_2, \cdots, u_{r-1}, u) \in \ker I_{p,2p+1}$. Applying σ_*, we find that $u = 0$. Note that $i^* i_* = -h$. By applying i^* to the equality

$$\sum_{j=1}^{r-1} i_* h^j \pi^* u_j = 0,$$

we get

$$\sum_{j=1}^{r-1} h^{j+1} \pi^* u_j = 0 \in L_{p-1} H_{k-2}(D).$$

The isomorphism in Proposition 1.3.1 again implies that $u_j = 0$ for $1 \leq j \leq r - 1$. This completes the proof for the case $k = 2p + 1$.

Now for $k \geq 2p + 2$, we reach the same situation as those in the case that $k = 2p$ or $k = 2p + 1$. More precisely, we give the complete argument by using mathematical induction.

Suppose that we have

$$
\begin{array}{ccccccccc}
\cdots \to & L_p H_{2p+m}(D) & \xrightarrow{i_*} & L_p H_{2p+m}(\tilde{X}_Y) & \xrightarrow{j^*} & L_p H_{2p+m}(U) & \xrightarrow{\delta_*} & 0 \\
& \downarrow \pi_* & & \downarrow \sigma_* & & \downarrow \cong & & \\
\cdots \to & L_p H_{2p+m}(Y) & \xrightarrow{(i_0)_*} & L_p H_{2p+m}(X) & \xrightarrow{j_0^*} & L_p H_{2p+m}(U) & \xrightarrow{(\delta_0)_*} & 0
\end{array}
\qquad (1.7)
$$

for some integer $m \geq 0$.

We want to prove that $I_{p,2p+m}$ is an isomorphism and

$$
\begin{array}{ccccccccc}
\cdots \to & L_p H_{2p+m+1}(D) & \xrightarrow{i_*} & L_p H_{2p+m+1}(\tilde{X}_Y) & \xrightarrow{j^*} & L_p H_{2p+m+1}(U) & \xrightarrow{\delta_*} & 0 \\
& \downarrow \pi_* & & \downarrow \sigma_* & & \downarrow \cong & & \\
\cdots \to & L_p H_{2p+m+1}(Y) & \xrightarrow{(i_0)_*} & L_p H_{2p+m+1}(X) & \xrightarrow{j_0^*} & L_p H_{2p+m+1}(U) & \xrightarrow{(\delta_0)_*} & 0
\end{array}
\qquad (1.8)
$$

Once this step is done, it completes the proof of the theorem.

From the assumption (1.7), we have

$$L_p H_{2p+m}(\tilde{X}_Y) = \sigma^* L_p H_{2p+m}(X) + \Sigma_{j=0}^{r-1} i_* h^j \pi^* L_{p-j} H_{2p+m-2j}(Y). \qquad (1.9)$$

19

From (1.4) for $k = 2p + m$ and (1.9), we obtain the surjectivity of $I_{p,2p+m}$ for the case that $k = 2p + m$.

To prove the injectivity, consider $(u_1, u_2, \cdots, u_{r-1}, u) \in \ker I_{p,2p+m}$. Applying σ_*, we find that $u = 0$. Note that $i^* i_* = -h$. By applying i^* to the equality

$$\sum_{j=1}^{r-1} i_* h^j \pi^* u_j = 0,$$

we get

$$\sum_{j=1}^{r-1} h^{j+1} \pi^* u_j = 0 \in L_{p-1} H_{k-2}(D).$$

The isomorphism in Proposition 1.3.1 once again implies that $u_j = 0$ for $1 \le j \le r-1$. This completes the proof for the case $k = 2p + m$. Now (1.7) automatically reduces to (1.8) and this completes the proof of the theorem.

□

As an application, this result gives many examples of smooth projective manifolds (even rational ones) for which the Griffiths group of p-cycles is infinitely generated (even modulo torsion) for $p \ge 2$. Recall that the Griffiths group $\mathrm{Griff}_p(X)$ is defined to be the p-cycles homologically equivalent to zero modulo the subgroup of p-cycles algebraically equivalent to zero.

Example: Note the fact in [F1] that $\mathrm{Griff}_2(\tilde{X}_Y) \cong L_2 H_4(\tilde{X}_Y)_{hom}$. For $X = \mathrm{P}^5$, $Y \subset \mathrm{P}^4$ the general hypersurface of degree 5, we obtain an infinite dimensional $\mathbb{Q}$-vector space $\mathrm{Griff}_2(\tilde{X}_Y) \otimes \mathbb{Q}$ from the fact $\dim_{\mathbb{Q}}(\mathrm{Griff}_1(Y) \otimes \mathbb{Q}) = \infty$ (cf. [Cl]). It gives the example mentioned in Remark 1.1.

From the blowup formula for Lawson homology and Clemens' result [Cl], we have the following

Corollary 1.3.1. *For each $n \ge 5$, there exists a rational manifold X with $\dim(X) = n$ such that*

$$\dim_{\mathbb{Q}} \left\{ \mathrm{Griff}_p(X) \otimes \mathbb{Q} \right\} = \infty, \quad 2 \le p \le n - 3.$$

Proof. Note that $\mathrm{Griff}_p(X) \cong L_p H_{2p}(X)_{hom}$ for any smooth projective variety X. Now the remaining argument is the direct result of Theorem 1.1.2 and the above result of Clemens [Cl].

□

More generally, from the blowup formula for Lawson homology and a result given in Chapter 6 ([H6]), we have the following

Corollary 1.3.2. *For any integers $p > 1$ and $k \geq 0$, there exists a* rational *projective manifold X such that $L_pH_{k+2p}(X) \otimes \mathbb{Q}$ is infinite dimensional vector space over $\mathbb{Q}$.*

Proof. It follows from the blowup formula for Lawson homology and Theorem 1.4 in [H6]. For example, if $p = 2$, $k = 1$, we can find a rational projective manifold X with $\dim(X) = 6$ such that $L_2H_5(X) \otimes \mathbb{Q}$ is infinite dimensional $\mathbb{Q}$-vector space. $\qquad\square$

1.4   The proof of the main theorem

The following result will be used several times in the proof of our main theorem:

Theorem 1.4.1. (Friedlander [F1]) *Let X be any smooth projective variety of dimension n. Then we have the following isomorphisms*

$$
\begin{cases}
L_{n-1}H_{2n}(X) \cong \mathbb{Z}, \\[4pt]
L_{n-1}H_{2n-1}(X) \cong H_{2n-1}(X, \mathbb{Z}), \\[4pt]
L_{n-1}H_{2n-2}(X) \cong H_{n-1,n-1}(X, \mathbb{Z}) = NS(X) \\[4pt]
L_{n-1}H_k(X) = 0 \quad for \quad k > 2n.
\end{cases}
$$

$\qquad\square$

Remark 1.4.1. *In the following, we adopt the notational convention $H_k(X) = H_k(X, \mathbb{Z})$.*

Now we begin the proof of our main results. There are two parts of the proof of the main theorem: $p = 1$ and $p = n - 2$.

Proof of the main theorem ($p = 1$):

Case A: $\sigma_* : L_1H_k(\tilde{X}_Y)_{hom} \to L_1H_k(X)_{hom}$ is injective.

We will use the commutative diagrams in Lemma 1.2.1–1.2.3.

Let $a \in L_1H_k(\tilde{X}_Y)_{hom}$ be such that $\sigma_*(a) = 0$. By Lemma 1.2.1, we have $j^*(a) = 0 \in L_1H_k(U)$ and hence there exists an element $b \in L_1H_k(D)$ such that $i_*(b) = a$. Set $\tilde{b} = \pi_*(b)$. By the commutative diagram in Lemma 1.2.1 again, we have $(i_0)_*(\tilde{b}) = 0 \in L_1H_k(X)$. By the exactness of the rows in the commutative diagram, there exists an element $\tilde{c} \in L_1H_{k+1}(U)$ such that the image of $\tilde{c}$ under the boundary map $(\delta_0)_* : L_1H_{k+1}(U) \to L_1H_k(Y)$ is $\tilde{b}$. Note that δ_* is the other boundary map $\delta_* : L_1H_{k+1}(U) \to L_1H_k(D)$. Therefore, $\pi_*(b - \delta_*(\tilde{c})) = 0 \in L_1H_k(Y)$ and $j_*(b - \delta_*(\tilde{c})) = a$. Now by

the "revised" Projective Bundle Theorem and Dold-Thom theorem ([DT]), we have
$L_1 H_k(D) \cong L_1 H_k(Y) \oplus L_0 H_{k-2}(Y) \oplus H_{k-4}(Y) \oplus \cdots \cong L_1 H_k(Y) \oplus H_{k-2}(Y) \oplus H_{k-4}(Y) \oplus \cdots$.
We know $b - \delta_*(\tilde{c}) \in H_{k-2}(Y) \oplus H_{k-4}(Y) \oplus \cdots$. By the explicit formula of the cohomology
(and homology) for a blowup ([GH], [V1]), we know each map $H_{k-2*}(Y) \to H_k(\tilde{X}_Y)$ is
injective. Hence a must be zero in $L_1 H_k(\tilde{X}_Y)$. This is the injectivity of σ_*.

Case B: $\sigma_* : L_1 H_k(\tilde{X}_Y)_{hom} \to L_1 H_k(X)_{hom}$ is surjective.

Let $a \in L_1 H_k(X)_{hom}$. From the surjectivity of the map $\sigma_* : L_1 H_k(\tilde{X}_Y) \to L_1 H_k(X)$,
there exists an element $\tilde{a} \in L_1 H_k(\tilde{X}_Y)$ such that $\sigma_*(\tilde{a}) = a$. Set $\tilde{b} = \Phi_{1,k}(\tilde{a})$. By the
commutative diagram in Lemma 1.2.1, we have $j^*(\tilde{b}) = 0 \in H_k^{BM}(U)$. From the exactness
of the rows of the diagram in Lemma 1.2.1, we have an element $\tilde{c} \in H_k(D)$ such that
$i_*(\tilde{c}) = \tilde{b}$. Set $c = \pi_*(\tilde{c})$. Then $(i_0)_*(c) = 0$ by the assumption of a and the commutative
of the diagram in Lemma 1.2.1. Using the exactness of rows in Lemma 1.2.1 again,
we can find an element $d \in H_{k+1}^{BM}(U)$ such that $(\delta_0)_*(d) = c$. Hence $i_*(\tilde{c} - \delta_*(d)) = \tilde{b} \in H_k(\tilde{X}_Y)$ and $\pi_*(\tilde{c} - \delta_*(d)) = 0$. Now we need to use the formula $L_1 H_k(D) \cong L_1 H_k(Y) \oplus H_{k-2}(Y) \oplus H_{k-4}(Y) \oplus \cdots$ again. From this we can find an element $e \in L_1 H_k(D)$
such that $\Phi_{1,k}(e) = \tilde{c} - \delta(d)$. Obviously, $\Phi_{1,k}(\tilde{a} - i_*(e)) = 0$ and $\sigma_*(\tilde{a} - i_*(e)) = a$ as we
want.

$\square$

Proof of the main theorem ($p = n - 2$):

Case 1: σ_* is injective.

The injectivity of

$$j_0^* : L_{n-2} H_k(X)_{hom} \to L_{n-2} H_k(U)_{hom}$$

is trivial since $\dim(Y) \leq n-2$, where $j_0 : U \to X$ is the inclusion. In fact, if $\dim(Y) < n-2$, then $j_0^* : L_{n-2} H_k(X) \to L_{n-2} H_k(U)$ is an isomorphism and so is $j_0^* : L_{n-2} H_k(X)_{hom} \to L_{n-2} H_k(U)_{hom}$. If $\dim(Y) = n - 2$, then for $k \geq 2(n-2) + 1$ the injectivity of j_0^* is
from the commutative diagram in Lemma 1.2.2, as well as the vanishing of $L_{n-2} H_k(Y)$
and $H_k(Y)$; for $k = 2(n - 2)$, the injectivity of j_0^* is from the commutative diagram
in Lemma 1.2.2, and the nontriviality of $(i_0)_* : H_{2(n-2)}(Y) \to H_{2(n-2)}(X)$, since Y is a
Kähler submanifold of X with complex dimension $n - 2$.

Now we need to prove $j^* : L_{n-2} H_k(\tilde{X}_Y)_{hom} \to L_{n-2} H_k(U)_{hom}$ is injective, where
$j : U \to \tilde{X}_Y$ the inclusion. Let $a \in L_{n-2} H_k(\tilde{X}_Y)_{hom}$ such that $j^*(a) = 0 \in L_{n-2} H_k(U)_{hom}$,
then there exists an element $b \in L_{n-2} H_k(D)$ such that $i_*(b) = a$. Now by the commutative
diagram in Corollary 1.2.1, we have $j_0^*(\sigma_*(a)) = 0$. Set $a' \equiv \sigma_*(a)$. From the exactness
of localization sequence in the bottom row in Corollary 1.2.1, there is an element $b' \in L_{n-2} H_k(Y)$ such that $(i_0)_*(b') = a'$.

Claim: In the commutative diagram in Corollary 1.2.1, there exists an element $c' \in L_{n-2} H_{k+1}(U)$ such that $(\delta_0)_*(c') = b'$ under the map $(\delta_0)_* : L_{n-2} H_{k+1}(U) \to L_{n-2} H_k(Y)$
and $\delta_*(c') = b$ under the map $\delta_* : L_{n-2} H_{k+1}(U) \to L_{n-2} H_k(D)$.

Proof of the claim: Since $\Phi_{n-2,k} : L_{n-2} H_k(Y) \cong H_k(Y)$ (note: $k \geq 2(n-2) \geq \dim(Y)$), we use the same notation b' for its image in $H_k(Y)$ since $L_{n-2}H_k(Y) \to H_k(Y)$ is injective for all $k \geq 2(n-2)$. At the beginning of the proof of the injectivity of the main theorem, we have already shown that $j_0^* : L_{n-2}H_k(X)_{hom} \to L_{n-2}H_k(U)_{hom}$ is injective. That is to say, $(i_0)_*(b') = 0 \in L_{n-2}H_k(X)_{hom}$. Hence there exists an element $c \in L_{n-2}H_{k+1}(U)$ such that whose image is b' under the boundary map $(\delta_0)_* : L_{n-2}H_{k+1}(U) \to L_{n-2}H_k(Y)$. Let $\tilde{b}$ be the image of c under the map $L_{n-2}H_{k+1}(U) \to L_{n-2}H_k(D)$. Now $\pi_*(\tilde{b} - b) = 0 \in L_{n-2}H_k(Y)$ and $i_*(\Phi_{n-2,k}(\tilde{b} - b)) = 0 \in H_k(\tilde{X}_Y)$, by Proposition 1.2.1, we have $\Phi_{n-2,k}(\tilde{b} - b) = 0$. Since $\Phi_{n-2,k}$ is injective on $L_{n-2}H_k(D)$ (see Theorem 1.4.1), we get $\tilde{b} - b = 0$. This c satisfies both conditions of the claim.

$\square$

Now everything is clear. The element a comes from the element c in $L_{n-2}H_{k+1}(U)$. By the exactness of the localization sequence in the upper row in Lemma 1.2.1, we get $a = 0 \in L_{n-2}H_k(\tilde{X}_Y)$. This completes the proof of the injectivity.

Case 2: σ_* is surjective.

Similar to the injectivity, the surjectivity of

$$j_0^* : L_{n-2}H_k(X)_{hom} \to L_{n-2}H_k(U)_{hom}$$

is trivial since the $\dim(Y) \leq n - 2$, where $j_0 : U \to X$ is the inclusion. In fact, if $\dim(Y) < n - 2$, $j_0^* : L_{n-2}H_k(X) \to L_{n-2}H_k(U)$ is an isomorphism and so is $j_0^* : L_{n-2}H_k(X)_{hom} \to L_{n-2}H_k(U)_{hom}$. If $\dim(Y) = n - 2$, then the surjectivity of j_0^* is from the commutative diagram in Lemma 1.2.3, and the isomorphism $\Phi_{n-2,2(n-2)} : L_{n-2}H_{2(n-2)}(Y) \cong H_{2(n-2)}(Y) \cong \mathbb{Z}$.

We only need to show $j^* : L_{n-2}H_k(\tilde{X}_Y)_{hom} \cong L_{n-2}H_k(U)_{hom}$, where $j : U \to \tilde{X}_Y$ the inclusion. There are a few cases.

(a) For the case that $k = 2(n - 2)$, the map $j^* : L_{n-2}H_k(\tilde{X}_Y) \to L_{n-2}H_k(U)$ is a surjective map. Hence the induced map j^* on $L_{n-2}H_k(\tilde{X}_Y)_{hom}$ is also surjective by trivial reason.

(b) The case that $k = 2(n - 2) + 1$. By the commutative diagram in Lemma 1.2.2, and note that the map $\Phi_{n-2,2(n-2)} : L_{n-2}H_{2(n-2)}(D) \to H_{2(n-2)}(D)$ is injective, we have, for $a \in L_{n-2}H_{2(n-2)+1}(U)_{hom}$, the image of a under the boundary map $\delta_* : L_{n-2}H_{2(n-2)+1}(U) \to L_{n-2}H_{2n}(D)$ must be zero. Hence a comes from an element $b \in L_{n-2}H_{2(n-2)+1}(\tilde{X}_Y)$. If $\bar{b} := \Phi_{n-2,2(n-2)+1}(b) \neq 0$, then $\exists c \in L_{n-2}H_{2(n-2)+1}(D)$ such that $b - i_*(c) \in L_{n-2}H_{2(n-2)+1}(\tilde{X}_Y)_{hom}$ and $j^*(b - i_*(c)) = a$. In fact, since $j^*(\bar{b}) = 0$, there exists $\bar{c} \in H_{2(n-2)+1}(D)$ such that $(i_0)_*(\bar{c}) = \bar{b}$. Note that $\Phi_{n-2,2(n-2)+1} : L_{n-2}H_{2(n-2)+1}(D) \to H_{2(n-2)+1}(D)$ is an isomorphism by Theorem 1.4.1), then there exists $c \in L_{n-2}H_{2(n-2)+1}(D)$ such that $\Phi_{n-2,2(n-2)+1}(c) = \bar{c}$. This shows the surjectivity in this case.

(c) Now we only need to consider the situation that $k \geq 2(n-2) + 2$. In this case, the surjectivity of $j^* : L_{n-2}H_k(\tilde{X}_Y)_{hom} \to L_{n-2}H_k(U)_{hom}$ is from the commutative diagram in Lemma 1.2.2, and the surjectivity of the map $\Phi_{n-2,k} : L_{n-2}H_k(D) \to H_k(D)$ (see Theorem 1.4.1)). In fact, if $a \in L_{n-2}H_k(U)_{hom}$, then by the exactness of the commutative diagram in Lemma 1.2.2, there is an element $b \in L_{n-2}H_k(\tilde{X}_Y)$ such that $j^*(b) = a$. Set $\bar{b} = \Phi_{n-2,k}(b)$. Since $j^*(\bar{b}) = 0 \in H_k^{BM}(U)$, $\exists \bar{c} \in H_k(D)$ such that $i_*(\bar{c}) = \bar{b}$. Now $\Phi_{n-2,k} : L_{n-2}H_k(D) \cong H_k(D)$ (See Theorem 1.4.1)), there exists $c \in L_{n-2}H_k(D)$ such that $\Phi_{n-2,k}(c) = \bar{c}$. The commutative diagram in Lemma 1.2.2 implies that $\Phi_{n-2,k}(b - i_*(c)) = 0$, i.e., $b - i_*(c) \in L_{n-2}H_k(\tilde{X}_Y)_{hom}$. The exactness of the upper row in Lemma 1.2.2 gives $j^*(b - i_*(c)) = a$. This completes the surjectivity in this case.

This completes the proof for a blow-up along a smooth subvariety Y of codimension at least 2 in X.

Now recall the weak factorization Theorem proved in [AKMW] (and also [Wl]) as follows:

Theorem 1.4.2. *([AKMW] Theorem 0.1.1, [Wl]) Let $\varphi \colon X \to X'$ be a birational map of smooth complete varieties over an algebraically closed field of characteristic zero, which is an isomorphism over an open set U. Then f can be factored as a sequence of birational maps*

$$ X = X_0 \overset{\varphi_1}{\dashrightarrow} X_1 \overset{\varphi_2}{\dashrightarrow} \cdots \overset{\varphi_{n+1}}{\dashrightarrow} X_n = X' $$

where each X_i is a smooth complete variety, and $\varphi_{i+1} : X_i \to X_{i+1}$ is either a blowing-up or a blowing-down of a smooth subvariety disjoint from U.

Note that $\varphi : X \to X'$ is birational between projective manifolds. We complete the proof of for the birational invariance of $L_{n-2}H_k(X)_{hom}$ for any smooth X by applying the above theorem. □

Remark 1.4.2. *Once we know birational invariance under blowup of Lawson homology groups $L_p H_k(X)_{hom}$, where p and k are given as in Theorem 1.1.1. Then we can deduce the result of any birational transformations without using the Weak Factorization Theorem. Instead, we need to use Hironaka desingularization theorem, together with the functoriality properties in section 1.2. The detail is given as follows. Let $\varphi : X \dashrightarrow X'$ be a birational*

24

map. Then there exist by the desingularization theorem (cf. [Hi1])

$$\hat{\varphi} : \hat{X} \rightarrow X', \tau : \hat{X} \rightarrow X,$$

where $\hat{\varphi}$ is a morphism and τ is the composition of a sequence of blowups along smooth centers. By using the desingularization theorem once again, we have

$$\psi : \hat{X}' \rightarrow \hat{X}, \tau' : \hat{X}' \rightarrow X',$$

where ψ is a morphism and τ' is the composition of a sequence of blow ups along smooth centers. Furthermore, ψ is the quasi-inverse of $\hat{\varphi}$ in the sense that $\tau' = \hat{\varphi} \circ \psi$.

Now we can define the homomorphism $\varphi_* : L_p H_k(X)_{hom} \rightarrow L_p H_k(X')_{hom}$ as $\hat{\varphi}_*$, by using the fact τ_* is an isomorphism from $L_p H_k(\hat{X})_{hom}$ to $L_p H_k(X)_{hom}$, as we proved in the first step. Now we prove that φ_* is an isomorphism of abelian groups.

Note that τ'_* is an isomorphism, we see that $\hat{\varphi}_*$ is surjective since $\tau'_* = \hat{\varphi}_* \circ \psi_*$ is surjective. Thus we proved the surjectivity of φ_* for birational maps. From this, now we prove the injectivity of φ_*. Note that from the definition, $\varphi_* = \hat{\varphi}_*$. Since the surjectivity holds for any birational map by the previous step, ψ_* is surjective. Hence it suffices to show that $\hat{\varphi}_* \circ \psi_*$ is injective. This is true since $\hat{\varphi}_* \circ \psi_* = \tau'_*$ is an isomorphism.

Remark 1.4.3. *Griffiths [G] showed the nontriviality of the Griffiths group of 1-cycles of general quintic hypersurfaces in $\mathbb{P}^4$ and Friedlander [F1] showed that $L_1 H_2(X)_{hom} \cong \mathrm{Griff}_1(X)$ for any smooth projective variety X. Hence, in general, this is a **nontrivial** birational invariant even for projective threefolds.*

Chapter 2

The Generalized Hodge Conjecture for 1-cycles and codimension two algebraic cycles

2.1 Introduction

In this chapter, all varieties are defined over $\mathbb{C}$. Let X be a smooth projective variety with dimension n. Let $\mathcal{Z}_p(X)$ be the space of algebraic p-cycles on X. Set $\mathcal{Z}^{n-p}(X) \equiv \mathcal{Z}_p(X)$. There is a natural map

$$cl_q : \mathcal{Z}^q(X) \to H^{2q}(X, \mathbb{Z})$$

called **the cycle class map**.

Tensoring with $\mathbb{Q}$, we have

$$cl_q \otimes \mathbb{Q} : \mathcal{Z}^q(X) \otimes \mathbb{Q} \to H^{2q}(X, \mathbb{Q}).$$

It is well known that $cl_q(\mathcal{Z}_q(X)) \subseteq H^{q,q}(X) \cap \rho(H^{2q}(X, \mathbb{Z}))$, where $\rho : H^{2q}(X, \mathbb{Z}) \to H^{2q}(X, \mathbb{C})$ is the coefficient homomorphism and $H^{q,q}(X)$ denotes the (q, q)-component in the Hodge decomposition (cf. [GH], [Lew1]).

There are known examples where $cl_q(\mathcal{Z}_q(X)) \neq H^{q,q}(X) \cap \rho(H^{2q}(X, \mathbb{Z}))$ (cf. [BCC] p.134-125], [Lew2]). We recall:

27

The Hodge Conjecture (for codimension-q cycles): *The rational cycle class map*

$$cl_q \otimes \mathbb{Q} : \mathcal{Z}^q(X) \otimes \mathbb{Q} \to H^{q,q}(X) \cap H^{2q}(X, \mathbb{Q})$$

is surjective.

The Hodge Conjecture over $\mathbb{Z}$: *The rational cycle class map*

$$cl_q : \mathcal{Z}^q(X) \to H^{q,q}(X) \cap \rho(H^{2q}(X, \mathbb{Z}))$$

is surjective.

We shall denote by Hodge^{q,q}(X, $\mathbb{Q}$) the statement that: " The Hodge Conjecture for codimension-q cycles is true for X". Similarly, we denote by Hodge^{q,q}(X, $\mathbb{Z}$) the corresponding statement for the Hodge Conjecture over $\mathbb{Z}$.

More generally, we can define a filtration on $H_k(X, \mathbb{Q})$ as follows:

Definition 2.1.1. ([FM],§7]) *Denote by* $\tilde{F}_p H_k(X, \mathbb{Q}) \subseteq H_k(X, \mathbb{Q})$ *the maximal sub-(Mixed) Hodge structure of span $k - 2p$. (See [Gro] and [FM].) The sub-$\mathbb{Q}$ vector spaces* $\tilde{F}_p H_k(X, \mathbb{Q})$ *form a decreasing filtration of sub-Hodge structures:*

$$\cdots \subseteq \tilde{F}_p H_k(X, \mathbb{Q}) \subseteq \tilde{F}_{p-1} H_k(X, \mathbb{Q}) \subseteq \cdots \subseteq \tilde{F}_0 H_k(X, \mathbb{Q}) \subseteq H_k(X, \mathbb{Q})$$

and $\tilde{F}_p H_k(X, \mathbb{Q})$ *vanishes if $2p > k$. This filtration is called the* **Hodge filtration**.

A homological version of the arithmetic filtration (see [[Lew1],§7]) is given in the following definition:

Definition 2.1.2. ([FM],§7]) *Denote by* $G_p H_k(X, \mathbb{Q}) \subseteq H_k(X, \mathbb{Q})$ *the $\mathbb{Q}$-vector subspace of $H_k(X, \mathbb{Q})$ generated by the images of mappings $H_k(Y, \mathbb{Q}) \to H_k(X, \mathbb{Q})$, induced from all morphisms $Y \to X$ of varieties of dimension$\leq k - p$. The subspaces $G_p H_k(X, \mathbb{Q})$ also form a decreasing filtration called the* **geometric filtration**:

$$\cdots \subseteq G_p H_k(X, \mathbb{Q}) \subseteq G_{p-1} H_k(X, \mathbb{Q}) \subseteq \cdots \subseteq G_0 H_k(X, \mathbb{Q}) \subseteq H_k(X, \mathbb{Q}).$$

Since X is smooth, the Weak Lefschetz Theorem implies that $G_0 H_k(X, \mathbb{Q}) = H_k(X, \mathbb{Q})$. Since $H_k(Y, \mathbb{Q})$ vanishes for k greater than twice the dimension of Y, $G_p H_k(X, \mathbb{Q})$ vanishes if $2p > k$.

It was proved in [Gro] that, for any smooth variety X, the geometric filtration is finer than the Hodge filtration, i.e., $G_p H_k(X, \mathbb{Q}) \subseteq \tilde{F}_p H_k(X, \mathbb{Q})$, for all p and k.

The Generalized Hodge Conjecture: For any smooth variety X,

$$G_p H_k(X, \mathbb{Q}) = \tilde{F}_p H_k(X, \mathbb{Q}) \tag{2.1}$$

for all p and k. Using the notation given in [Lew1], we denote by $\widetilde{GHC}(p, k, X)$ the assertion that the equation (2.1) is true.

Definition 2.1.3. *The **Lawson homology** $L_p H_k(X)$ of p-cycles is defined by*

$$L_p H_k(X) = \pi_{k-2p}(\mathcal{Z}_p(X)) \quad for \quad k \geq 2p \geq 0,$$

where $\mathcal{Z}_p(X)$ is provided with a natural topology (cf. [F1], [L1]). For general background, the reader is referred to Lawson' survey paper [L2].

There are two special cases.

(a) If $p = 0$, then for all $k \geq 0$, $L_0 H_k(X) \cong H_k(X, \mathbb{Z})$ by Dold-Thom Theorem [DT].

(b) If $k = 2p$, then $L_p H_{2p}(X) = \mathcal{Z}_p(X)/\mathcal{Z}_p(X)_{alg}$, where $\mathcal{Z}_p(X)_{alg}$ denotes the algebraic p-cycles on X which are algebraic equivalent to zero.

In [FM], Friedlander and Mazur showed that there are natural maps, called **cycle class maps**

$$\Phi_{p,k} : L_p H_k(X) \to H_k(X, \mathbb{Z}).$$

Define

$$L_p H_k(X)_{hom} := \ker\{\Phi_{p,k} : L_p H_k(X) \to H_k(X, \mathbb{Z})\}.$$

$$T_p H_k(X) := \mathrm{Im}\{\Phi_{p,k} : L_p H_k(X) \to H_k(X, \mathbb{Z})\}$$

and

$$T_p H_k(X, \mathbb{Q}) := T_p H_k(X) \otimes \mathbb{Q}.$$

It was proved in [[FM],§7] that, for any smooth variety X, $T_p H_k(X, \mathbb{Q}) \subseteq G_p H_k(X, \mathbb{Q})$ for all p and k. Hence

29

$$T_p H_k(X, \mathbb{Q}) \subseteq G_p H_k(X, \mathbb{Q}) \subseteq \tilde{F}_p H_k(X, \mathbb{Q}). \tag{2.2}$$

In this chapter, we will use the tools in Lawson homology and the methods given in [H1] to show the following main result:

Theorem 2.1.1. *Let X be a smooth projective variety. If the Hodge conjecture for codimension 2 cycles over $\mathbb{Z}$ holds for X, i.e., if we have $\mathrm{Hodge}^{2,2}(X, \mathbb{Z})$, then it holds for any smooth projective variety X' birational to X. That is, $\mathrm{Hodge}^{2,2}(X, \mathbb{Z})$ is a birationally invariant assertion for smooth varieties X.*

Remark 2.1.1. *The above theorem remains true if $\mathbb{Z}$ is replaced by $\mathbb{Q}$. Since $\mathrm{Hodge}^{2,2}(X, \mathbb{Q})$ implies $\mathrm{Hodge}^{n-2,n-2}(X, \mathbb{Q})$ for $n \geq 4$ (cf. [[Lew1], p.91]), $\mathrm{Hodge}^{n-2,n-2}(X, \mathbb{Q})$ is also a birationally invariant property of smooth n-dimensional varieties X.*

As a corollary, we have

Corollary 2.1.1. *If X is a rational manifold with $\dim(X) \leq 5$, then the Hodge conjecture $\mathrm{Hodge}^{p,p}(X, \mathbb{Q})$ is true for $1 \leq p \leq \dim(X)$. In fact, $\mathrm{Hodge}^{p,p}(X', \mathbf{Z})$ is true except possibly for $p = 3, \dim(X) = 5$.*

Remark 2.1.2. *By using the technique of the diagonal decomposition, Bloch and Srinivas [BS] showed that, for a smooth projective variety X, $\mathrm{Hodge}^{2,2}(X, \mathbb{Q})$ holds if the Chow group of 0-cycles $\mathrm{Cho}(X) \cong \mathbb{Z}$. Laterveer [Lat] generalized this technique and showed the Hodge Conjecture holds for a class of projective manifolds with small chow groups.*

Corollary 2.1.2. *Let X be a smooth projective variety of dimension ≤ 5 such that the Hodge Conjecture is known to be true, i.e., $\mathrm{Hodge}^{p,p}(X, \mathbb{Q})$ holds for all p. Then the Hodge Conjecture holds for all smooth projective varieties X' which are birationally equivalent to X. Non-rational examples of such an X include general abelian varieties or the product of at most five elliptic curves. For more examples, the reader is referred to the survey book [Lew1].*

Our second main result is the following

Theorem 2.1.2. *The assertion $\widetilde{GHC}(n - 2, k, X)$ is a birationally invariant property of smooth n-dimensional varieties X when $k \geq 2(n-2)$. More precisely, if $\widetilde{GHC}(n-2, k, X)$ holds for a smooth variety X, then $\widetilde{GHC}(n - 2, k, X')$ holds for any smooth variety X' birational to X.*

We also show that

Proposition 2.1.1. *The assertion that "$T_{n-2} H_k(X, \mathbb{Q}) = \tilde{F}_{n-2} H_k(X, \mathbb{Q})$ holds" is a birationally invariant property of smooth n-dimensional varieties X when $k \geq 2(n - 2)$.*

Similarly, for 1-cycles, we can show the following.

Proposition 2.1.2. *For integer $k \geq 2$, the assertion that "$T_1 H_k(X, \mathbb{Q}) = \tilde{F}_1 H_k(X, \mathbb{Q})$ holds" is a birationally invariant property of smooth n-dimensional varieties X.*

and

Theorem 2.1.3. *For any integer $k \geq 2$, the assertion $\widetilde{GHC}(1, k, X)$ is a birationally invariant property of smooth varieties X.*

Remark 2.1.3. *For the case $k = \dim(X)$, Lewis has already obtained this result in [Lew1].*

Corollary 2.1.3. *For any smooth rational variety X with $\dim(X) \leq 4$, the Generalized Hodge Conjecture holds.*

The main tools used to prove this result are: the long exact localization sequence given by Lima-Filho in [Li1], the explicit formula for Lawson homology of codimension-one cycles on a smooth projective manifold given by Friedlander in [F1], and the weak factorization theorem proved by Wlodarczyk in [Wl] and in [AKMW].

2.2 The proof of the main theorems

Let X be a smooth projective manifold of dimension n. In the following, we will denote by $H_{p,q}(X)$ the image of $H^{n-p,n-q}(X)$ under the Poincare duality isomorphism $H^{2n-p-q}(X, \mathbb{C}) \cong H_{p+q}(X, \mathbb{C})$.

Let X be a smooth projective manifold and $i_0 : Y \hookrightarrow X$ be a smooth subvariety of codimension r. Let $\sigma : \tilde{X}_Y \to X$ be the blowup of X along Y, $i : D := \sigma^{-1}(Y) \hookrightarrow \tilde{X}_Y$ the exceptional divisor of the blowing up, and $\pi : D \to Y$ the restriction of σ to D. Set $U := X - Y \cong \tilde{X}_Y - D$. Denote by j_0 the inclusion $U \subset X$ and j the inclusion $U \subset \tilde{X}_Y$.

Now I list the Lemmas and Corollaries given in [H1].

Lemma 2.2.1. *For each p, we have the following commutative diagram*

$$
\begin{array}{ccccccccc}
\cdots \to & L_p H_k(D) & \xrightarrow{i_*} & L_p H_k(\tilde{X}_Y) & \xrightarrow{j^*} & L_p H_k(U) & \xrightarrow{\delta_*} & L_p H_{k-1}(D) & \to & \cdots \\
& \downarrow \pi_* & & \downarrow \sigma_* & & \downarrow \cong & & \downarrow \pi_* & & \\
\cdots \to & L_p H_k(Y) & \xrightarrow{(i_0)_*} & L_p H_k(X) & \xrightarrow{j_0^*} & L_p H_k(U) & \xrightarrow{(\delta_0)_*} & L_p H_{k-1}(Y) & \to & \cdots
\end{array}
$$

Remark 2.2.1. *Since π_* is surjective (this follows from the explicit formula for the Lawson homology of D, i.e., the Projective Bundle Theorem in [FG]), it is easy to see that σ_* is surjective.*

31

Corollary 2.2.1. *If $p = 0$, then we have the following commutative diagram*

$$\cdots \to H_k(D) \xrightarrow{i_*} H_k(\tilde{X}_Y) \xrightarrow{j^*} H_k^{BM}(U) \xrightarrow{\delta_*} H_{k-1}(D) \to \cdots$$
$$\downarrow \pi_* \qquad\qquad \downarrow \sigma_* \qquad\qquad \downarrow \cong \qquad\qquad \downarrow \pi_*$$
$$\cdots \to H_k(Y) \xrightarrow{(i_0)^*} H_k(X) \xrightarrow{j_0^*} H_k^{BM}(U) \xrightarrow{(\delta_0)^*} H_{k-1}(Y) \to \cdots$$

Moreover, if $x \in H_k(D)$ vanishes under π_* and i_*, then $x = 0 \in H_k(D)$.

Corollary 2.2.2. *If $p = n - 2$, then we have the following commutative diagram*

$$\to L_{n-2}H_k(D) \xrightarrow{i_*} L_{n-2}H_k(\tilde{X}_Y) \xrightarrow{j^*} L_{n-2}H_k(U) \xrightarrow{\delta_*} L_{n-2}H_{k-1}(D) \to \cdots$$
$$\downarrow \pi_* \qquad\qquad \downarrow \sigma_* \qquad\qquad \downarrow \cong \qquad\qquad \downarrow \pi_*$$
$$\to L_{n-2}H_k(Y) \xrightarrow{(i_0)^*} L_{n-2}H_k(X) \xrightarrow{j_0^*} L_{n-2}H_k(U) \xrightarrow{(\delta_0)^*} L_{n-2}H_{k-1}(Y) \to \cdots$$

Lemma 2.2.2. *For each p, we have the following commutative diagram*

$$\cdots \to L_pH_k(D) \xrightarrow{i_*} L_pH_k(\tilde{X}_Y) \xrightarrow{j^*} L_pH_k(U) \xrightarrow{\delta_*} L_pH_{k-1}(D) \to \cdots$$
$$\downarrow \Phi_{p,k} \qquad\qquad \downarrow \Phi_{p,k} \qquad\qquad \downarrow \Phi_{p,k} \qquad\qquad \downarrow \Phi_{p,k-1}$$
$$\cdots \to H_k(D) \xrightarrow{i_*} H_k(\tilde{X}_Y) \xrightarrow{j^*} H_k^{BM}(U) \xrightarrow{\delta_*} H_{k-1}(D) \to \cdots$$

In particular, it is true for $p = 1, n - 2$.

Lemma 2.2.3. *For each p, we have the following commutative diagram*

$$\cdots \to L_pH_k(Y) \xrightarrow{(i_0)^*} L_pH_k(X) \xrightarrow{j^*} L_pH_k(U) \xrightarrow{(\delta_0)^*} L_pH_{k-1}(Y) \to \cdots$$
$$\downarrow \Phi_{p,k} \qquad\qquad \downarrow \Phi_{p,k} \qquad\qquad \downarrow \Phi_{p,k} \qquad\qquad \downarrow \Phi_{p,k-1}$$
$$\cdots \to H_k(Y) \xrightarrow{(i_0)^*} H_k(X) \xrightarrow{j^*} H_k^{BM}(U) \xrightarrow{(\delta_0)^*} H_{k-1}(Y) \to \cdots$$

In particular, it is true for $p = 1, n - 2$.

Remark 2.2.2. *All the commutative diagrams of long exact sequences remain commutative and exact after tensoring with $\mathbb{Q}$. We will use these Lemmas and corollaries with rational coefficients.*

The following result proved by Friedlander will be used several times:

Theorem 2.2.1. (Friedlander [F1]) *Let X be any smooth projective variety of dimension n. Then we have the following isomorphisms*

$$\begin{cases} L_{n-1}H_{2n}(X) \cong \mathbb{Z}, \\ L_{n-1}H_{2n-1}(X) \cong H_{2n-1}(X,\mathbb{Z}), \\ L_{n-1}H_{2n-2}(X) \cong H_{n-1,n-1}(X,\mathbb{Z}) = NS(X) \\ L_{n-1}H_k(X) = 0 \quad for \quad k > 2n. \end{cases}$$

where $NS(X)$ is the Néron-Severi group of X.

2.2.1 The proof of Theorem 2.1.1 for a blowup

In what follows we drop reference to the coefficient homomorphism ρ, and denote by $H_k(X,\mathbb{Z})$ its image in $H_k(X,\mathbb{C})$.

There are two cases to consider:

Case 1: If $\Phi_{n-2,2(n-2)} : L_{n-2}H_{2(n-2)}(X) \to H_{2(n-2)}(X,\mathbb{Z}) \cap H_{n-2,n-2}(X)$ is surjective, we will show that $\Phi_{n-2,2(n-2)} : L_{n-2}H_{2(n-2)}(\tilde{X}_Y) \to H_{2(n-2)}(\tilde{X}_Y,\mathbb{Z}) \cap H_{n-2,n-2}(\tilde{X}_Y)$ is also surjective.

Let $b \in H_{n-2,n-2}(\tilde{X}_Y) \cap H_{2(n-2)}(\tilde{X}_Y,\mathbb{Z})$. Set $a \equiv \sigma_*(b) \in H_{2(n-2)}(X,\mathbb{Z})$. Since σ_* preserves the type, we have $a \in H_{2(n-2)}(X,\mathbb{Z}) \cap H_{n-2,n-2}(X)$. Now by assumption, there exists an element $\tilde{a} \in L_{n-2}H_{2(n-2)}(X)$ such that

$$\Phi_{n-2,2(n-2)}(\tilde{a}) = a.$$

Now since $\sigma_* : L_{n-2}H_{2(n-2)}(\tilde{X}_Y) \to L_{n-2}H_{2(n-2)}(X)$ is surjective, there exists an element $\tilde{b} \in L_{n-2}H_{2(n-2)}(\tilde{X}_Y)$ such that $\sigma_*(\tilde{b}) = \tilde{a}$. Now $\Phi_{n-2,2(n-2)}(\tilde{b}) - b$ is mapped to zero under σ_* on $H_{2(n-2)}(\tilde{X}_Y,\mathbb{Z})$. By the commutative diagram in the long exact sequences in Corollary 2.2.1, there exists an element $c \in H_{2(n-2)}(D,\mathbb{Z})$ such that $i_*(c) = \Phi_{n-2,2(n-2)}(\tilde{b}) - b$. Using Corollary 2.2.1 once again, we have $\pi_*(c) = 0$. This follows from the fact that $\dim(Y) = n - r \le n - 2$ and hence $(i_0)_* : H_{2(n-2)}(Y,\mathbb{Z}) \to H_{2(n-2)}(X,\mathbb{Z})$ is injective. From the blowup formula for the singular homology, $i_*|_{\ker \pi_*}$ is injective. Now by assumption, b and $\tilde{b}$ are non-torsion elements. Hence c is not a torsion element in $H_{2(n-2)}(D,\mathbb{Z})$, i.e., $c \in H_{2(n-2)}(D,\mathbb{Z})_{\text{free}}$, the torsion free part of $H_{2(n-2)}(D,\mathbb{Z})$.

Since i_* preserves the type, we have the following

Claim: $c \in H_{2(n-2)}(D,\mathbb{Z}) \cap H_{n-2,n-2}(D)$.

Proof. Note that

$$H_{2(n-2)}(D, \mathbb{Z})_{\text{free}} \subset H_{2(n-2)}(D, \mathbb{C}) = H_{n-2,n-2}(D) \oplus H_{n-1,n-3}(D) \oplus H_{n-3,n-1}(D).$$

Now $c = c_0 + c_1 + \bar{c}_1 \in H_{2(n-2)}(D, \mathbb{C})$ such that $c_0 \in H_{n-2,n-2}(D)$, $c_1 \in H_{n-1,n-3}(D)$ and hence $\bar{c}_1 \in H_{n-3,n-1}(D)$. Note that the complexification of i_* is the map $i_* \otimes \mathbb{C} :$ $H_{2(n-2)}(D, \mathbb{C}) \to H_{2(n-2)}(\tilde{X}_Y, \mathbb{C})$. If $i_* \otimes \mathbb{C}(c_1) = 0$, we have $c_1 = 0$. In fact, $i_* \otimes \mathbb{C}(c_1) = 0$ and the exactness of the long exact sequence in the upper row in Corollary 2.2.1 implies that an element $d \in H^{BM}_{2(n-2)+1}(U, \mathbb{C})$ such that $\delta_*(d) = c_1$. We use the commutative diagram in Corollary 2.2.1 again. From the commutativity of the diagram in Corollary 2.2.1, we have the image of d under the boundary map $(\delta_0)_*$ must zero in $H_{2(n-2)}(Y, \mathbb{C})$. This follows from the fact that the complex dimension of $\dim(Y) \leq n-2$ and the Hodge type of d is of type $(n-1, n-3)$. Now by the exactness of the long exact sequence in the lower row in Corollary 2.2.1, there exists an element $e \in H_{2(n-2)+1}(X, \mathbb{C})$ such that $j_0^*(e) = d$. It is well-known that $\sigma_* : H_{2(n-2)+1}(\tilde{X}_Y, \mathbb{C}) \to H_{2(n-2)+1}(X, \mathbb{C})$ is surjective. Therefore, there exists $\tilde{e} \in H_{2(n-2)+1}(\tilde{X}_Y, \mathbb{C})$ such that $\sigma_*(\tilde{e}) = e$. We get $d = j^*(\tilde{e})$ and hence $c_1 = 0 \in H_{2(n-2)}(D, \mathbb{C})$ by the exactness of the the upper row sequence in Corollary 2.2.1. This implies $\bar{c}_1 = 0$ and hence $c \in H_{n-2,n-2}(D)$. This finishes the proof of the claim.

$\square$

Since $\dim(D) = n-1$, hence by Theorem 2.2.1, the map

$$\Phi_{n-2,2(n-2)} : L_{n-2} H_{2(n-2)}(D) \to H_{2(n-2)}(D, \mathbb{Z}) \cap H_{n-2,n-2}(D)$$

is an isomorphism. Set $\tilde{c} \equiv \Phi_{n-2,2(n-2)}(c)$. Therefore, $\Phi_{n-2,2(n-2)}\{\tilde{b} - i_*(\tilde{c})\} = b$. Hence $\Phi_{n-2,2(n-2)} : L_{n-2} H_{2(n-2)}(\tilde{X}_Y) \to H_{2(n-2)}(\tilde{X}_Y, \mathbb{Z}) \cap H_{n-2,n-2}(\tilde{X}_Y)$ is surjective .

On the other hand, we need to show

Case 2: If $\Phi_{n-2,2(n-2)} : L_{n-2} H_{2(n-2)}(\tilde{X}_Y) \to H_{2(n-2)}(\tilde{X}_Y, \mathbb{Z}) \cap H_{n-2,n-2}(\tilde{X}_Y)$ is surjective, then $\Phi_{n-2,2(n-2)} : L_{n-2} H_{2(n-2)}(X) \to H_{2(n-2)}(X, \mathbb{Z}) \cap H_{n-2,n-2}(X)$ is also surjective.

This part is relatively easy. Let $a \in H_{2(n-2)}(X) \cap H_{n-2,n-2}(X)$. Since

$$\sigma_* : H_{2(n-2)}(\tilde{X}_Y, \mathbb{Z}) \to H_{2(n-2)}(X, \mathbb{Z})$$

is surjective and $\sigma_* \otimes \mathbb{C} : H_{2(n-2)}(\tilde{X}_Y, \mathbb{C}) \to H_{2(n-2)}(X, \mathbb{C})$ preserves the Hodge type, there exists an element $b \in H_{2(n-2)}(\tilde{X}_Y, \mathbb{Z}) \cap H_{n-2,n-2}(\tilde{X}_Y)$ such that $\sigma_*(b) = a$. Now by assumption, we have an element $\tilde{b} \in L_{n-2} H_{2(n-2)}(\tilde{X}_Y)$ such that $\Phi_{n-2,2(n-2)}(\tilde{b}) = b$. Set $\tilde{a} \equiv \sigma_*(\tilde{b})$. Then from the commutative of the diagram, we have $\Phi_{n-2,2(n-2)}(\tilde{a}) = a$. This is exactly the surjectivity in this case.

This completes the proof for a blowup along a smooth codimension at least two subvariety Y in X.

$\square$

2.2.2 The proof of Theorem 2.1.2 for a blowup

Now we have the following:

Proposition 2.2.1. *The assertion that "$T_{n-2}H_k(X,\mathbb{Q}) = \tilde{F}_{n-2}H_k(X,\mathbb{Q})$ holds" is a birationally invariant property of smooth n-dimensional varieties X when $k \geq 2(n-2)$.*

Proof. There are two cases to consider:

Case A: If $\Phi_{n-2,k} : L_{n-2}H_k(X) \otimes \mathbb{Q} \to \tilde{F}_{n-2}H_k(X,\mathbb{Q})$ is surjective, we want to show $\Phi_{n-2,k} : L_{n-2}H_k(\tilde{X}_Y) \otimes \mathbb{Q} \to \tilde{F}_{n-2}H_k(\tilde{X}_Y,\mathbb{Q})$ is also surjective.

Let $a \in \tilde{F}_{n-2}H_k(\tilde{X}_Y,\mathbb{Q})$, set $b = \sigma_*(a) \in \tilde{F}_{n-2}H_k(X,\mathbb{Q})$. By assumption, there exists $\tilde{b} \in L_{n-2}H_k(X,\mathbb{Q})$ such that $\Phi_{n-2,k}(\tilde{b}) = b$. By the blowup formula in Lawson homology (see [H1]), we know that $\sigma_* : L_{n-2}H_k(\tilde{X}_Y,\mathbb{Q}) \to L_{n-2}H_k(X,\mathbb{Q})$ is surjective, there exists an element $\tilde{a} \in L_{n-2}H_k(\tilde{X}_Y,\mathbb{Q})$ such that $\sigma_*(\tilde{a}) = \tilde{b}$. By the commutative diagram in Lemma 2.2.1 and Corollary 2.2.1, we have $j^*(\Phi_{n-2,k}(\tilde{a}) - a) = 0 \in H_k^{BM}(U,\mathbb{Q})$. The exactness of the localization sequence in the rows in Corollary 2.2.1 implies that there exists an element $c \in H_k(D,\mathbb{Q})$ such that $i_*(c) = \Phi_{n-2,k}(\tilde{a}) - a$. Since the $\dim(D) = n-1$ and D is smooth, by Theorem 2.1, we know the natural transformation $\Phi_{n-2,k} : L_{n-2}H_k(D) \to H_k(D)$ is an isomorphism for $k \geq 2(n-2)+1$. Hence $\Phi_{n-2,k} : L_{n-2}H_k(D) \otimes \mathbb{Q} \cong H_k(D,\mathbb{Q})$. Therefore there exists $\tilde{c} \in L_{n-2}H_k(D) \otimes \mathbb{Q}$ such that $\Phi_{n-2,k}(\tilde{c}) = c$. Now it is obvious that $\Phi_{n-2,k}(\tilde{a} - i_*(\tilde{c})) = a$. The proof of the case $k = 2(n-p)$ is from the proof of Theorem 2.1.1. This is the surjectivity as we want.

Case B: If $\Phi_{n-2,k} : L_{n-2}H_k(\tilde{X}_Y) \otimes \mathbb{Q} \to \tilde{F}_{n-2}H_k(\tilde{X}_Y,\mathbb{Q})$ is surjective, we want to show $\Phi_{n-2,k} : L_{n-2}H_k(X) \otimes \mathbb{Q}) \to \tilde{F}_{n-2}H_k(X,\mathbb{Q})$ is also surjective. We can use an argument similar to the **Case 2** above. Suppose $b \in \tilde{F}_{n-2}H_k(X,\mathbb{Q})$. Then there exists a $\tilde{b} \in \tilde{F}_{n-2}H_k(\tilde{X}_Y,\mathbb{Q})$ such that $\sigma_*(\tilde{b}) = b$ by the blowup formula for the singular homology with $\mathbb{Q}$-coefficients. By assumption, there exists an $\tilde{a} \in L_{n-2}H_k(\tilde{X}_Y) \otimes \mathbb{Q}$ such that $\Phi_{n-2,k}(\tilde{a}) = \tilde{b}$. Set $a = \sigma_*(\tilde{a})$. Then $a \in L_{n-2}H_k(X) \otimes \mathbb{Q}$ and $\Phi_{n-2,k}(a) = b$. This finishes the proof of the surjectivity in this case.

$\square$

Now we give the proof of Theorem 2.1.2. First, we suppose that

$$G_{n-2}H_k(X,\mathbb{Q}) = \tilde{F}_{n-2}H_k(X,\mathbb{Q}).$$

We will show

$$G_{n-2}H_k(\tilde{X}_Y, \mathbb{Q}) = \tilde{F}_{n-2}H_k(\tilde{X}_Y, \mathbb{Q})$$

case by case.

For $k > 2n$, $\tilde{F}_{n-2}H_k(\tilde{X}_Y) = 0$ and hence nothing needs to be proved.

For $k = 2n$, $G_{n-2}H_k(\tilde{X}_Y) = \tilde{F}_{n-2}H_k(\tilde{X}_Y) = \mathbb{Z}$, so the result is true.

For $k = 2n - 1, 2n - 2$, $G_{n-2}H_k(\tilde{X}_Y) = \tilde{F}_{n-2}H_k(\tilde{X}_Y) = H_k(\tilde{X}_Y, \mathbb{Q})$ follows from the definitions of the geometric filtration and the Hodge filtration.

The only case left is $k = 2n - 3$ since the case that $k = 2n - 4$ has been proved in Theorem 2.1.1. In this case, $T_{n-2}H_k(M, \mathbb{Q}) = G_{n-2}H_k(M, \mathbb{Q})$ has been proved in [H3] for any smooth projective variety M. The assumption $G_{n-2}H_k(X, \mathbb{Q}) = \tilde{F}_{n-2}H_k(X, \mathbb{Q})$ is equivalent to $T_{n-2}H_k(X, \mathbb{Q}) = \tilde{F}_{n-2}H_k(X, \mathbb{Q})$ in this situation. Hence $T_{n-2}H_k(\tilde{X}_Y, \mathbb{Q}) = \tilde{F}_{n-2}H_k(\tilde{X}_Y, \mathbb{Q})$ follows from Proposition 2.2.1. Now by (2.2), we have $G_{n-2}H_k(\tilde{X}_Y, \mathbb{Q}) = \tilde{F}_{n-2}H_k(\tilde{X}_Y, \mathbb{Q})$.

On the other hand, it has been proved in [[Lew1], Lemma 13.6] that $G_{n-2}H_k(X, \mathbb{Q}) \cong \tilde{F}_{n-2}H_k(X, \mathbb{Q})$ holds if $G_{n-2}H_k(\tilde{X}_Y, \mathbb{Q}) \cong \tilde{F}_{n-2}H_k(\tilde{X}_Y, \mathbb{Q})$. The last part is exactly the assumption. This finishes the proof of Theorem 2.1.2 for one blowup over a smooth subvariety of codimension at least two.

$\square$

2.2.3   The proof of Theorem 2.1.3 for a blowup

Similarly, for 1-cycles, we have the following.

Proposition 2.2.2. *For integer $k \geq 2$, the assertion that "$T_1H_k(X, \mathbb{Q}) = \tilde{F}_1H_k(X, \mathbb{Q})$ holds" is a birationally invariant property of smooth n-dimensional varieties X.*

Proof. As before, there are two cases to consider:

Case a: If $T_1H_k(X, \mathbb{Q}) = \tilde{F}_1H_k(X, \mathbb{Q})$ holds, then

$$T_1H_k(\tilde{X}_Y, \mathbb{Q}) = \tilde{F}_1H_k(\tilde{X}_Y, \mathbb{Q})$$

holds. By the theorems in [[FM],§7], $T_1H_k(M, \mathbb{Q}) \subseteq \tilde{F}_1H_k(M, \mathbb{Q})$ holds for any smooth variety M. We only need to show $T_1H_k(\tilde{X}_Y, \mathbb{Q}) \supseteq \tilde{F}_1H_k(\tilde{X}_Y, \mathbb{Q})$. The argument is similar to the proof of the Theorem 1.3 in [H3]. I give the detail as follows:

Let $a \in \tilde{F}_1H_k(\tilde{X}_Y, \mathbb{Q})$, set $b = \sigma_*(a) \in \tilde{F}_1H_k(X, \mathbb{Q})$. By assumption, there exists $\tilde{b} \in L_1H_k(X, \mathbb{Q})$ such that $\Phi_{1,k}(\tilde{b}) = b$. By the blowup formula in Lawson homology (see [H1]), we know that $\sigma_* : L_1H_k(\tilde{X}_Y, \mathbb{Q}) \to L_1H_k(X, \mathbb{Q})$ is surjective, there exists an element $\tilde{a} \in L_1H_k(\tilde{X}_Y, \mathbb{Q})$ such that $\sigma_*(\tilde{a}) = \tilde{b}$. By the commutative diagram in Lemma 2.2.1 and Corollary 2.2.1, we have $j^*(\Phi_{1,k}(\tilde{a}) - a) = 0 \in H_k^{BM}(U, \mathbb{Q})$. The exactness of the localization sequence in the rows in Corollary 2.2.1 implies that there exists an element $c \in H_k(D, \mathbb{Q})$ such that $i_*(c) = \Phi_{1,k}(\tilde{a}) - a$. Set $d = \pi_*(c) \in L_1 H_k(Y) \otimes \mathbb{Q}$. By the commutative diagram in Corollary 2.2.1, d maps to zero under $(i_0)_* : H_k(Y, \mathbb{Q}) \to H_k(X, \mathbb{Q})$. Hence there exists an element $e \in H_{k+1}^{BM}(U, \mathbb{Q})$ such that $(\delta_0)_*(e) = d$. Let $\tilde{d} \in H_k(D, \mathbb{Q})$ be the image of e under this boundary map $\delta_* : H_{k+1}^{BM}(U, \mathbb{Q}) \to H_k(D, \mathbb{Q})$, i.e., $\tilde{d} = \delta_*(e)$. Therefore, the image of $c - \tilde{d}$ is zero in $H_k(Y, \mathbb{Q})$ under π_* and is zero in $H_k(\tilde{X}_Y, \mathbb{Q})$ under i_*. By the blowup formula in Lawson homology (see [H1]), we know such an element $c - \tilde{d}$ in the image of some $f \in L_1 H_k(D) \otimes \mathbb{Q}$, i.e., $\Phi_{1,k}(f) = c - \tilde{d}$. Hence we get $\Phi_{1,k}(\tilde{a} - i_*(f)) = a$. This is the surjectivity as we want.

Case b: If $T_1 H_k(\tilde{X}_Y, \mathbb{Q}) = \tilde{F}_1 H_k(\tilde{X}_Y, \mathbb{Q})$ holds, then $T_1 H_k(X, \mathbb{Q}) = \tilde{F}_1 H_k(X, \mathbb{Q})$ holds. This part is relatively easy. As before, we only need to show $T_1 H_k(X, \mathbb{Q}) \supseteq \tilde{F}_1 H_k(X, \mathbb{Q})$.

Let $b \in \tilde{F}_1 H_k(X, \mathbb{Q})$. Since $\sigma : \tilde{X}_Y \to X$ is the blowup along the smooth variety Y, we have $\sigma_*(\tilde{F}_1 H_k(\tilde{X}_Y, \mathbb{Q})) \subseteq \tilde{F}_1 H_k(X, \mathbb{Q})$. In fact, the inclusion is an equality. (See [Lew2] Lemma.13.6) Therefore, there is an element $a \in \tilde{F}_1 H_k(\tilde{X}_Y, \mathbb{Q})$ such that $\sigma_*(a) = b$. By assumption, there is an element $\tilde{a} \in L_1 H_k(\tilde{X}_Y, \mathbb{Q})$ such that $\Phi_{1,k}(\tilde{a}) = a$. Set $\tilde{b} = \Phi_{1,k}(\tilde{a}) \in L_1 H_k(X, \mathbb{Q})$. By the naturality of $\Phi_{1,k}$, we have $\sigma_*(\tilde{b}) = b$. This is the surjectivity as we need.

$\square$

Now we give the proof of Theorem 2.1.3. First, suppose $G_1 H_k(X, \mathbb{Q}) = \tilde{F}_1 H_k(X, \mathbb{Q})$. We want to show that $G_1 H_k(\tilde{X}_Y, \mathbb{Q}) = \tilde{F}_1 H_k(\tilde{X}_Y, \mathbb{Q})$.

Now comparing the blowup formula for Lawson homology (cf. [H1]) and for singular homology (both with $\mathbb{Q}$ coefficients) along the same smooth subvariety Y of codimension at least two, we find the same new components, i.e.,

$$\bigoplus_{j=1}^{r-1} H_{k-2j}(Y, \mathbb{Q}),$$

both in $L_1 H_k(\tilde{X}_Y, \mathbb{Q})$ and $H_k(\tilde{X}_Y, \mathbb{Q})$.

This, together with (2.2), implies that the new component of this blowup along Y in $G_1 H_k(\tilde{X}_Y, \mathbb{Q})$ contains $\bigoplus_{j=1}^{r-1} H_{k-2j}(Y, \mathbb{Q})$. Since $G_1 H_k(\tilde{X}_Y, \mathbb{Q}) \subseteq H_k(\tilde{X}_Y, \mathbb{Q})$, the new component of this blowup along Y in $G_1 H_k(\tilde{X}_Y, \mathbb{Q})$ is also contained in $\bigoplus_{j=1}^{r-1} H_{k-2j}(Y, \mathbb{Q})$. Therefore

$$G_1 H_k(\tilde{X}_Y, \mathbb{Q}) \cong \left\{ \bigoplus_{j=1}^{r-1} H_{k-2j}(Y, \mathbb{Q}) \right\} \bigoplus G_1 H_k(X, \mathbb{Q}) \tag{2.3}$$

Similarly,

$$\tilde{F}_1 H_k(\tilde{X}_Y, \mathbb{Q}) \cong \left\{ \bigoplus_{j=1}^{r-1} H_{k-2j}(Y, \mathbb{Q}) \right\} \bigoplus \tilde{F}_1 H_k(X, \mathbb{Q}) \tag{2.4}$$

From (2.3) and (2.4), we deduce that $G_1 H_k(\tilde{X}_Y, \mathbb{Q}) = \tilde{F}_1 H_k(\tilde{X}_Y, \mathbb{Q})$.

On the other hand, we also need to show that if $G_1 H_k(\tilde{X}_Y, \mathbb{Q}) = \tilde{F}_1 H_k(\tilde{X}_Y, \mathbb{Q})$, then $G_1 H_k(X, \mathbb{Q}) = \tilde{F}_1 H_k(X, \mathbb{Q})$. An argument similar to the one given in **Case B** works. Lewis [[Lew1], Lemma 13.6] proved this part in a more general setting.

This finishes the proof of Theorem 2.1.3 for a blowup along a smooth subvariety with codimension at least two.

$\square$

Now recall the weak factorization Theorem proved in [AKMW] (and also [Wl]) as follows:

Theorem 2.2.2. *([AKMW] Theorem 0.1.1, [Wl]) Let $f : X \to X'$ be a birational map of smooth complete varieties over an algebraically closed field of characteristic zero, which is an isomorphism over an open set U. Then f can be factored as*

$$X = X_0 \overset{\varphi_1}{\dashrightarrow} X_1 \overset{\varphi_2}{\dashrightarrow} \cdots \overset{\varphi_{n+1}}{\dashrightarrow} X_n = X'$$

where each X_i is a smooth complete variety, and $\varphi_{i+1} : X_i \to X_{i+1}$ is either a blowing-up or a blowing-down of a smooth subvariety disjoint from U.

Moreover, if $X - U$ and $X' - U$ are simple normal crossings divisors, then the same is true for each $X_i - U$, and the center of the blowing-up has normal crossings with each $X_i - U$.

Hence $\text{Hodge}^{2,2}(X, \mathbb{Q})$, $\widetilde{GHC}(n-2, k, X)$ and $\widetilde{GHC}(1, k, X)$ are birationally invariant properties about the smooth manifold X.

$\square$

The proof of the Corollary 2.1.1 and 2.1.2 are based on Theorem 2.1.1, Remark 1.1 and the strong Lefschetz Theorem. By using the strong Lefschetz Theorem, one can show that $\text{Hodge}^{p,p}(X, \mathbb{Q}) \Rightarrow \text{Hodge}^{n-p,n-p}(X, \mathbb{Q})$ for $2p \leq n$. (cf. [Lew1] for the details.)

$\square$

The Corollary 2.1.3 is obvious from Theorem 2.1.2 and Theorem 2.1.3.

$\square$

2.3   A remark on generalizations

From the proof of the Theorem 2.1.1 and 2.1.2, we can draw the following conclusions:

(a) Fix $n > 0$ and $0 \le p \le n$. If we have Hodge$^{i,i}(Y, \mathbb{Q})$ for all $i \le p$ and all smooth projective variety Y, i.e., the Hodge conjecture is true for every smooth projective variety Y with $\dim(Y) = n$ and for algebraic cycles with codimension $\le p$, then Hodge$^{p+1,p+1}(X, \mathbb{Q})$ is a birational invariant statement for every smooth projective X with $\dim(X) \le n + 2$. For example, if we have Hodge$^{2,2}(Y, \mathbb{Q})$ for all 4-folds Y, then Hodge$^{p,p}(X, \mathbb{Q})$ is a birational statement for any integer $0 \le p \le \dim(X)$ and smooth projective varieties X with $\dim(X) \le 7$.

For the Generalized Hodge Conjecture, we have

(b) Fix $n > 0$ and $0 \le p \le n$. If we have $\widetilde{GHC}(i, k, Y)$ for $i \le p$, i.e., the Generalized Hodge Conjecture is true for every smooth projective Y with $\dim(Y) = n$ and for algebraic cycles with codimension $\le p$, then $\widetilde{GHC}(m - p - 1, k, X)$ is a birational invariant statement for every smooth projective variety X with $\dim(X) = m \le n+2$.

Similarly,

(c) Fix $n > 0$ and $0 \le p \le n$. If we have $\widetilde{GHC}(i, k, Y)$ for $i \le p$, i.e., the Generalized Hodge Conjecture is true for every smooth projective Y with $\dim(Y) = n$ and for algebraic cycles with dimension $\le p$, then $\widetilde{GHC}(p+1, k, X)$ is a birational invariant statement for every smooth projective variety X with $\dim(X) = m \le n + 2$.

As a corollary of part (b) and (c), we have, for example, if we have $\widetilde{GHC}(1, 3, Y)$ for all 3-folds Y, then $\widetilde{GHC}(p, k, X)$ is a birational statement for X with $\dim(X) \le 5$.

Chapter 3

Some relations between topological and geometric filtrations on smooth projective manifolds

3.1 Introduction

In this chapter, all varieties are defined over $\mathbb{C}$. Let X be a projective variety with dimension n. Let $\mathcal{Z}_p(X)$ be the space of algebraic p-cycles.

The **Lawson homology** $L_pH_k(X)$ of p-cycles is defined by

$$L_pH_k(X) = \pi_{k-2p}(\mathcal{Z}_p(X)) \quad for \quad k \geq 2p \geq 0,$$

where $\mathcal{Z}_p(X)$ is provided with a natural topology (cf. [F1], [L1]). For general background, the reader is referred to Lawson' survey paper [L2].

In [FM], Friedlander and Mazur showed that there are natural maps, called **cycle class maps**

$$\Phi_{p,k} : L_pH_k(X) \to H_k(X).$$

Definition 3.1.1.

$$L_pH_k(X)_{hom} := \ker\{\Phi_{p,k} : L_pH_k(X) \to H_k(X)\};$$

$$T_pH_k(X) := \mathrm{Image}\{\Phi_{p,k} : L_pH_k(X) \to H_k(X)\};$$

$$T_pH_k(X, \mathbb{Q}) := T_pH_k(X) \otimes \mathbb{Q}.$$

41

It was shown in [[FM], §7] that the subspaces $T_pH_k(X, \mathbb{Q})$ form a decreasing filtration:

$$\cdots \subseteq T_pH_k(X, \mathbb{Q}) \subseteq T_{p-1}H_k(X, \mathbb{Q}) \subseteq \cdots \subseteq T_0H_k(X, \mathbb{Q}) = H_k(X, \mathbb{Q})$$

and $T_pH_k(X, \mathbb{Q})$ vanishes if $2p > k$.

Definition 3.1.2. ([FM]) *Denote by $G_pH_k(X, \mathbb{Q}) \subseteq H_k(X, \mathbb{Q})$ the $\mathbb{Q}$-vector subspace of $H_k(X, \mathbb{Q})$ generated by the images of mappings $H_k(Y, \mathbb{Q}) \to H_k(X, \mathbb{Q})$, induced from all morphisms $Y \to X$ of varieties of dimension $\leq k - p$.*

*The subspaces $G_pH_k(X, \mathbb{Q})$ also form a decreasing filtration (called **geometric filtration**):*

$$\cdots \subseteq G_pH_k(X, \mathbb{Q}) \subseteq G_{p-1}H_k(X, \mathbb{Q}) \subset \cdots \subseteq G_0H_k(X, \mathbb{Q}) \subseteq H_k(X, \mathbb{Q})$$

If X is smooth, the Weak Lefschetz Theorem implies that $G_0H_k(X, \mathbb{Q}) = H_k(X, \mathbb{Q})$. Since $H_k(Y, \mathbb{Q})$ vanishes for k greater than twice the dimension of Y, $G_pH_k(X, \mathbb{Q})$ vanishes if $2p > k$.

The following results have been proved by Friedlander and Mazur in [FM]:

Theorem 3.1.1. ([FM]) *Let X be any projective variety.*

1. *For non-negative integers p and k,*

$$T_pH_k(X, \mathbb{Q}) \subseteq G_pH_k(X, \mathbb{Q}).$$

2. *When $k = 2p$,*

$$T_pH_{2p}(X, \mathbb{Q}) = G_pH_{2p}(X, \mathbb{Q}).$$

Question ([FM], [L2]): Does one have equality in Theorem 3.1.1 when X is a smooth projective variety?

Friedlander [F2] has the following result:

42

Theorem 3.1.2. ([F2]) *Let X be a smooth projective variety of dimension n. Assume that Grothendieck's Standard Conjecture B ([Gro]) is valid for a resolution of singularities of each irreducible subvariety of $Y \subset X$ of dimension $k - p$, then*

$$T_p H_k(X, \mathbb{Q}) = G_p H_k(X, \mathbb{Q}).$$

Remark 3.1.1. ([Lew1],§15.32) *The Grothendieck's Standard Conjecture B is known to hold for a smooth projective variety X in the following cases:*

1. $\dim X \leq 2$.

2. Flag manifolds X.

3. Smooth complete intersections X.

4. Abelian varieties (due to D. Lieberman [Lieb]).

In this chapter, we will use the tools in Lawson homology and the methods given in [H1] to show the following main results:

Theorem 3.1.3. *Let X be a smooth projective variety of dimension n. If the conclusion in Theorem 3.1.2 holds (without the assumption of Grothendieck's Standard Conjecture B) for X with $p = 1$, (resp. $p = n - 2$) (k arbitrary), then it also holds for any smooth projective variety X' which is birationally equivalent to X with $p = 1$, (resp. $p = n - 2$).*

Theorem 3.1.4. *For any smooth projective variety X,*

$$T_p H_{2p+1}(X, \mathbb{Q}) = G_p H_{2p+1}(X, \mathbb{Q}).$$

As corollaries, we have

Corollary 3.1.1. *Let X be a smooth projective 3-fold. We have $T_p H_k(X, \mathbb{Q}) = G_p H_k(X, \mathbb{Q})$ for all $k \geq 2p \geq 0$ except for the case $p = 1, k = 4$.*

43

Corollary 3.1.2. *Let X be a smooth projective 3-fold with $H^{2,0}(X) = 0$. Then*

$$T_p H_k(X, \mathbb{Q}) = G_p H_k(X, \mathbb{Q})$$

for any $k \geq 2p \geq 0$. In particular, it holds for X a smooth hypersurface and a complete intersection of dimension 3.

By using the Künneth formula in homology with rational coefficient, we have

Corollary 3.1.3. *Let X be the product of a smooth projective curve and a smooth simply connected projective surface. Then $T_p H_k(X, \mathbb{Q}) = G_p H_k(X, \mathbb{Q})$ for any $k \geq 2p \geq 0$.*

Corollary 3.1.4. *For 4-folds X, the assertion that $T_p H_k(X, \mathbb{Q}) = G_p H_k(X, \mathbb{Q})$ holds for all $k \geq 2p \geq 0$ is a birational invariant statement. In particular, if X is a rational manifold with $\dim(X) \leq 4$, then the conclusion in Theorem 3.1.2 holds for any $k \geq 2p \geq 0$ without assumption of Grothendieck's Standard Conjecture B .*

Remark 3.1.2. *A Conjecture given by Suslin (see [FHW], §7) implies that*

$$L_p H_{n+p}(X^n) \cong H_{n+p}(X^n).$$

As an application of Theorem 3.1.4 and Proposition 3.3.1, we have the following result:

Corollary 3.1.5. *If the Suslin Conjecture holds, then the topological filtration is the same as the geometric filtration for a smooth projective variety.*

The main tools to prove this result are: the long exact localization sequence given by Lima-Filho in [Li1], the explicit formula for Lawson homology of codimension-one cycles on a smooth projective manifold given by Friedlander in [F1], (and its generalization to general irreducible varieties, see below), and the weak factorization theorem proved by Wlodarczyk in [Wl] and in [AKMW].

3.2 The proof of the Theorem 3.1.3

Let X be a smooth projective manifold of dimension n and $i_0 : Y \hookrightarrow X$ be a smooth subvariety of codimension $r \geq 2$. Let $\sigma : \tilde{X}_Y \to X$ be the blowup of X along Y, $\pi : D = \sigma^{-1}(Y) \to Y$ the nature map, and $i : D = \sigma^{-1}(Y) \hookrightarrow \tilde{X}_Y$ the exceptional divisor of the blowup. Set $U := X - Y \cong \tilde{X}_Y - D$. Denote by j_0 the inclusion $U \subset X$ and j the inclusion $U \subset \tilde{X}_Y$.

Now I list the Lemmas and Corollaries given in [H1].

Lemma 3.2.1. *For each $p \geq 0$, we have the following commutative diagram*

$$
\begin{array}{ccccccccc}
\cdots \to & L_p H_k(D) & \xrightarrow{i_*} & L_p H_k(\tilde{X}_Y) & \xrightarrow{j^*} & L_p H_k(U) & \xrightarrow{\delta_*} & L_p H_{k-1}(D) & \to \cdots \\
& \downarrow \pi_* & & \downarrow \sigma_* & & \downarrow \cong & & \downarrow \pi_* & \\
\cdots \to & L_p H_k(Y) & \xrightarrow{(i_0)^*} & L_p H_k(X) & \xrightarrow{j_0^*} & L_p H_k(U) & \xrightarrow{(\delta_0)_*} & L_p H_{k-1}(Y) & \to \cdots
\end{array}
$$

Remark 3.2.1. *Since π_* is surjective (there is an explicitly formula for the Lawson homology of D, i.e., the Projective Bundle Theorem proved by Friedlander and Gabber, see [FG]), it is easy to see that σ_* is surjective.*

Corollary 3.2.1. *If $p = 0$, then we have the following commutative diagram*

$$
\begin{array}{ccccccccc}
\cdots \to & H_k(D) & \xrightarrow{i_*} & H_k(\tilde{X}_Y) & \xrightarrow{j^*} & H_k^{BM}(U) & \xrightarrow{\delta_*} & H_{k-1}(D) & \to \cdots \\
& \downarrow \pi_* & & \downarrow \sigma_* & & \downarrow \cong & & \downarrow \pi_* & \\
\cdots \to & H_k(Y) & \xrightarrow{(i_0)^*} & H_k(X) & \xrightarrow{j_0^*} & H_k^{BM}(U) & \xrightarrow{(\delta_0)_*} & H_{k-1}(Y) & \to \cdots
\end{array}
$$

Moreover, if $x \in H_k(D)$ maps to zero under π_* and i_*, then $x = 0 \in H_k(D)$.

Corollary 3.2.2. *If $p = n - 2$, then we have the following commutative diagram*

$$
\begin{array}{ccccccccc}
\to & L_{n-2} H_k(D) & \xrightarrow{i_*} & L_{n-2} H_k(\tilde{X}_Y) & \xrightarrow{j^*} & L_{n-2} H_k(U) & \xrightarrow{\delta_*} & L_{n-2} H_{k-1}(D) & \to \cdots \\
& \downarrow \pi_* & & \downarrow \sigma_* & & \downarrow \cong & & \downarrow \pi_* & \\
\to & L_{n-2} H_k(Y) & \xrightarrow{(i_0)^*} & L_{n-2} H_k(X) & \xrightarrow{j_0^*} & L_{n-2} H_k(U) & \xrightarrow{(\delta_0)_*} & L_{n-2} H_{k-1}(Y) & \to \cdots
\end{array}
$$

Lemma 3.2.2. *For each $p \geq 0$, we have the following commutative diagram*

$$
\begin{array}{ccccccccc}
\cdots \to & L_p H_k(D) & \xrightarrow{i_*} & L_p H_k(\tilde{X}_Y) & \xrightarrow{j^*} & L_p H_k(U) & \xrightarrow{\delta_*} & L_p H_{k-1}(D) & \to \cdots \\
& \downarrow \Phi_{p,k} & & \downarrow \Phi_{p,k} & & \downarrow \Phi_{p,k} & & \downarrow \Phi_{p,k-1} & \\
\cdots \to & H_k(D) & \xrightarrow{i_*} & H_k(\tilde{X}_Y) & \xrightarrow{j^*} & H_k^{BM}(U) & \xrightarrow{\delta_*} & H_{k-1}(D) & \to \cdots
\end{array}
$$

In particular, it is true for $p = 1, n - 2$.

Proof. See [Li1] and also [FM].

□

Lemma 3.2.3. *For each $p \geq 0$, we have the following commutative diagram*

$$
\begin{array}{ccccccccc}
\cdots \to & L_p H_k(Y) & \stackrel{(i_0)_*}{\to} & L_p H_k(X) & \stackrel{j^*}{\to} & L_p H_k(U) & \stackrel{(\delta_0)_*}{\to} & L_p H_{k-1}(Y) & \to \cdots \\
& \downarrow \Phi_{p,k} & & \downarrow \Phi_{p,k} & & \downarrow \Phi_{p,k} & & \downarrow \Phi_{p,k-1} & \\
\cdots \to & H_k(Y) & \stackrel{(i_0)_*}{\to} & H_k(X) & \stackrel{j^*}{\to} & H_k^{BM}(U) & \stackrel{(\delta_0)_*}{\to} & H_{k-1}(Y) & \to \cdots
\end{array}
$$

In particular, it is true for $p = 1, n - 2$.

Proof. See [Li1] and also [FM].

□

Remark 3.2.2. *The smoothness of X and Y is not necessary in the Lemma 3.2.3.*

Remark 3.2.3. *All the commutative diagrams of long exact sequences above remain commutative and exact when tensored with $\mathbb{Q}$. We will use these Lemmas and Corollaries with rational coefficients.*

The following result will be used several times in the proof of our main theorem:

Theorem 3.2.1. *(Friedlander [F1]) Let W be any smooth projective variety of dimension n. Then we have the following isomorphisms*

$$
\begin{cases}
L_{n-1} H_{2n}(W) \cong \mathbb{Z}, \\
L_{n-1} H_{2n-1}(W) \cong H_{2n-1}(X, \mathbb{Z}), \\
L_{n-1} H_{2n-2}(W) \cong H_{n-1,n-1}(X, \mathbb{Z}) = NS(W) \\
L_{n-1} H_k(X) = 0 \quad for \quad k > 2n.
\end{cases}
$$

The proof of Theorem 3.1.3 ($p = n - 2$):

There are two cases:

Case 1. If $T_p H_k(X, \mathbb{Q}) = G_p H_k(X, \mathbb{Q})$, then $T_p H_k(\tilde{X}_Y, \mathbb{Q}) = G_p H_k(\tilde{X}_Y, \mathbb{Q})$.

The injectivity of $T_p H_k(\tilde{X}_Y, \mathbb{Q}) \to G_p H_k(\tilde{X}_Y, \mathbb{Q})$ has been proved by Friedlander and Mazur in [FM]. We only need to show the surjectivity. Note that the case for $k = 2p + 1$ holds for any smooth projective variety (Theorem 3.1.4). We only need to consider the cases where $k \geq 2p + 2$. In these cases, $k - p \geq p + 2 = n$, from the definition of the geometric filtrations, we have $G_p H_k(\tilde{X}, \mathbb{Q}) = H_k(\tilde{X}_Y, \mathbb{Q})$ and $G_p H_k(X, \mathbb{Q}) = H_k(X, \mathbb{Q})$.

Let $b \in G_p H_k(\tilde{X}_Y, \mathbb{Q})$, and a be the image of b under the the map

$$\sigma_* : H_k(\tilde{X}_Y, \mathbb{Q}) \to H_k(X, \mathbb{Q}),$$

i.e., $\sigma_*(b) = a$. By assumption, there exists an element $\tilde{a} \in L_{n-2} H_k(X) \otimes \mathbb{Q}$ such that $\Phi_{n-2,k}(\tilde{a}) = a$. Since $\sigma_* : L_{n-2} H_k(\tilde{X}_Y) \otimes \mathbb{Q} \to L_{n-2} H_k(X) \otimes \mathbb{Q}$ is surjective ([H1]), there exists an element $\tilde{b} \in L_{n-2} H_k(X) \otimes \mathbb{Q}$ such that $\sigma_*(\tilde{b}) = \tilde{a}$. By the following commutative diagram

$$
\begin{array}{ccc}
L_{n-2} H_k(\tilde{X}_Y) \otimes \mathbb{Q} & \xrightarrow{\sigma_*} & L_{n-2} H_k(X) \otimes \mathbb{Q} \\
\downarrow \Phi_{n-2,k} & & \downarrow \Phi_{n-2,k} \\
H_k(\tilde{X}_Y, \mathbb{Q}) & \xrightarrow{\sigma_*} & H_k(X, \mathbb{Q}),
\end{array}
$$

we have $\Phi_{n-2,k}(\tilde{b}) - b$ maps to zero in $H_k(X, \mathbb{Q})$. By the commutative diagram in Corollary 3.2.1, $j^*(\Phi_{n-2,k}(\tilde{b}) - b) = 0 \in H_k^{BM}(U, \mathbb{Q})$. From the exactness of the upper long exact sequence in Corollary 3.2.1, there exists an element $c \in H_k(D, \mathbb{Q})$ such that $i_*(c) = \Phi_{n-2,k}(\tilde{b}) - b$. From Theorem 3.2.1, we find that $\Phi_{n-2,k} : L_{n-2} H_k(D) \otimes \mathbb{Q} \to H_k(D) \otimes \mathbb{Q}$ is an isomorphism for $k \geq 2n - 2$. Hence there exists an element $\tilde{c} \in L_{n-2} H_k(D) \otimes \mathbb{Q}$ such that $i_*(\Phi_{n-2,k}(\tilde{c})) = \Phi_{n-2,k}(\tilde{b}) - b$. Therefore $\Phi_{n-2,k}(\tilde{b} - i_*(\tilde{c})) = b$, i.e., the surjectivity of $T_p H_k(\tilde{X}_Y, \mathbb{Q}) \to G_p H_k(\tilde{X}_Y, \mathbb{Q})$.

On the other hand, we need to show

Case 2. If $T_p H_k(\tilde{X}_Y, \mathbb{Q}) = G_p H_k(\tilde{X}_Y, \mathbb{Q})$, then $T_p H_k(X, \mathbb{Q}) = G_p H_k(X, \mathbb{Q})$.

This part is relatively easy. By Theorem 3.1.4, we only need to consider the cases that $k \geq 2p + 2 = 2n - 2$. Let $a \in G_p H_k(X, \mathbb{Q}) = H_k(X, \mathbb{Q})$. From the blow up formula for singular homology (cf. [GH], [V1]), we know $\sigma_* : H_k(\tilde{X}_Y, \mathbb{Q}) \to H_k(X, \mathbb{Q})$ is surjective. Then there exists an element $b \in H_k(\tilde{X}_Y, \mathbb{Q})$ such that $\sigma_*(b) = a$. By assumption, we can find an element $\tilde{b} \in L_{n-2} H_k(\tilde{X}_Y, \mathbb{Q})$ such that $\Phi_{n-2,k}(\tilde{b}) = b$. Set $\tilde{a} = \sigma_*(\tilde{b})$. Then $\Phi_{n-2,k}(\tilde{a}) = a$ under the natural map $\Phi_{n-2,k}$. This is exactly the surjectivity we want.

This completes the proof for a blow-up along a smooth codimension at least two subvariety Y in X.

$\square$

The proof of Theorem 3.1.3 ($p = 1$):

The injectivity of the map $T_1 H_k(W, \mathbb{Q}) \to G_1 H_k(W, \mathbb{Q})$ has been proved for any smooth projective variety W by Friedlander and Mazur in [FM]. We only need to show the surjectivity under certain assumption.

Similar to the case $p = n - 2$, we also have two cases:

Case A. If $T_1 H_k(X, \mathbb{Q}) = G_1 H_k(X, \mathbb{Q})$, then $T_1 H_k(\tilde{X}_Y, \mathbb{Q}) = G_1 H_k(\tilde{X}_Y, \mathbb{Q})$.

From Theorem 3.1.4, the case where $k = 3$ holds for any smooth projective variety. We only need to consider the cases where $k \geq 4$.

Let $b \in G_1 H_k(\tilde{X}_Y, \mathbb{Q})$. Denote by a the image of b under the the map $\sigma_* : H_k(\tilde{X}_Y, \mathbf{Q}) \to H_k(X, \mathbb{Q})$, i.e., $\sigma_*(b) = a$. From the blow up formula for singular homology and the definition of the geometric filtration, we have $\sigma_*(G_1 H_k(\tilde{X}_Y, \mathbb{Q})) = G_1 H_k(X, \mathbb{Q})$.

By assumption, there exists an element $\tilde{a} \in L_1 H_k(X) \otimes \mathbb{Q}$ such that $\Phi_{1,k}(\tilde{a}) = a$. Since $\sigma_* : L_1 H_k(\tilde{X}_Y) \otimes \mathbb{Q} \to L_1 H_k(X) \otimes \mathbb{Q}$ is surjective ([H1]), there exists an element $\tilde{b} \in L_1 H_k(\tilde{X}_Y) \otimes \mathbb{Q}$ such that $\sigma_*(\tilde{b}) = \tilde{a}$. By the following commutative diagram

$$
\begin{array}{ccc}
L_1 H_k(\tilde{X}_Y) \otimes \mathbb{Q} & \overset{\sigma_*}{\to} & L_1 H_k(X) \otimes \mathbb{Q} \\
\downarrow \Phi_{1,k} & & \downarrow \Phi_{1,k} \\
H_k(\tilde{X}_Y, \mathbb{Q}) & \overset{\sigma_*}{\to} & H_k(X, \mathbb{Q}),
\end{array}
$$

we have $\Phi_{1,k}(\tilde{b}) - b$ maps to zero in $H_k(X, \mathbb{Q})$. By the commutative diagram in Corollary 3.2.1, $j^*(\Phi_{1,k}(\tilde{b}) - b) = 0 \in H_k^{BM}(U, \mathbb{Q})$. From the exactness of the upper long exact sequence in Corollary 3.2.1, there exists an element $c \in H_k(D, \mathbb{Q})$ such that $i_*(c) = \Phi_{1,k}(\tilde{b}) - b$. Set $d = \pi_*(c) \in H_k(Y, \mathbb{Q})$. By the commutative diagram in Corollary 3.2.1, d maps to zero under $(i_0)_* : H_k(Y, \mathbb{Q}) \to H_k(X, \mathbb{Q})$. Hence there exists an element $e \in H_{k+1}^{BM}(U, \mathbb{Q})$ such that whose image is d under the boundary map $(\delta_0)_*$. Let $\tilde{d} \in H_k(D, \mathbb{Q})$ be the image of e under this boundary map $\delta_* : H_{k+1}^{BM}(U, \mathbb{Q}) \to H_k(D, \mathbb{Q})$. Therefore, the image of $c - \tilde{d}$ is zero under π_* in $H_k(Y, \mathbb{Q})$ and is also zero under i_* in $H_k(\tilde{X}_Y, \mathbb{Q})$. Note that D is a bundle over Y with projective spaces as fibers. From the "projective bundle theorem" for the singular homology (cf.[GH]), we have $H_k(D, \mathbb{Q}) \cong H_k(Y, \mathbb{Q}) \oplus H_{k-2}(Y, \mathbb{Q}) \oplus \cdots \oplus H_{k-2r+2}(Y, \mathbb{Q})$. From this, we have $c - \tilde{d} \in H_{k-2}(Y, \mathbb{Q}) \oplus \cdots \oplus H_{k-2r+2}(Y, \mathbb{Q})$. By the revised Projective Bundle Theorem ([FG], and [H1] the revised case essentially due to Complex Suspension Theorem [L1]) and Dold-Thom Theorem [DT], we have $L_1 H_k(D, \mathbb{Q}) \cong L_1 H_k(Y, \mathbb{Q}) \oplus L_0 H_{k-2}(Y, \mathbb{Q}) \oplus \cdots \oplus L_{2-r} H_{k-2r+2}(Y, \mathbb{Q}) \cong L_1 H_k(Y, \mathbb{Q}) \oplus H_{k-2}(Y, \mathbb{Q}) \oplus \cdots \oplus H_{k-2r+2}(Y, \mathbb{Q})$, where r is the codimension of Y. Since $c - \tilde{d} \in H_{k-2}(Y, \mathbb{Q}) \oplus \cdots \oplus H_{k-2r+2}(Y, \mathbb{Q})$ and $L_0 H_{k-2}(Y, \mathbb{Q}) \oplus \cdots \oplus L_{2-r} H_{k-2r+2}(Y, \mathbb{Q}) \cong H_{k-2}(Y, \mathbb{Q}) \oplus \cdots \oplus H_{k-2r+2}(Y, \mathbb{Q})$, there exists an element $f \in L_1 H_k(D, \mathbb{Q})$ such that $\Phi_{1,k}(f) = c - \tilde{d}$. Therefore we obtain $\Phi_{1,k}(\tilde{b} - i_*(f)) = b$. This is the surjectivity we need.

Case B. If $T_1 H_k(\tilde{X}_Y, \mathbb{Q}) = G_1 H_k(\tilde{X}_Y, \mathbb{Q})$, then $T_1 H_k(X, \mathbb{Q}) = G_1 H_k(X, \mathbb{Q})$.

This part is also relatively easy. Note that $k \geq 4$. Let $a \in G_1 H_k(X, \mathbb{Q}) \subseteq H_k(X, \mathbb{Q})$, then there exists an element $b \in G_1 H_k(\tilde{X}_Y, \mathbb{Q})$ such that $\sigma_*(b) = a$. By assumption, we can find an element $\tilde{b} \in L_1 H_k(\tilde{X}_Y, \mathbb{Q})$ such that $\Phi_{1,k}(\tilde{b}) = b$. Set $\tilde{a} = \sigma_*(\tilde{b})$. Then $\Phi_{1,k}(\tilde{a}) = a$ under the natural transformation $\Phi_{1,k}$. This is exactly the surjectivity in these cases.

This completes the proof for one blow-up along a smooth codimension at least two subvariety Y in X.

$\square$

Now recall the weak factorization Theorem proved in [AKMW] (and also [Wl]) as follows:

Theorem 3.2.2. *([AKMW] Theorem 0.1.1, [Wl]) Let $\varphi\colon X \to X'$ be a birational map of smooth complete varieties over an algebraically closed field of characteristic zero, which is an isomorphism over an open set U. Then f can be factored as a sequence of birational maps*

$$X = X_0 \overset{\varphi_1}{\to} X_1 \overset{\varphi_2}{\to} \cdots \overset{\varphi_{n+1}}{\to} X_n = X'$$

where each X_i is a smooth complete variety, and $\varphi_{i+1}\colon X_i \to X_{i+1}$ is either a blowing-up or a blowing-down of a smooth subvariety disjoint from U.

$\square$

Remark 3.2.4. *From the proof of the Theorem 4.1.3, we can draw the following conclusions:*

1. *If*

$$T_r H_k(Y,\mathbb{Q}) = G_r H_k(Y,\mathbb{Q})$$

   for all k is true for algebraic r-cycles with $r \geq p$ for $\dim(Y) = n$, then

   $$\text{``}T_{p-1} H_k(X,\mathbb{Q}) = G_{p-1} H_k(X,\mathbb{Q}), \quad \forall k\text{''}$$

   is a birationally invariant statement for smooth projective varieties X with $\dim(X) \leq n + 2$.

2. *If*

$$T_r H_k(Y,\mathbb{Q}) = G_r H_k(Y,\mathbb{Q})$$

   for all k is true for r-algebraic cycles with $r \leq p$ for $\dim(Y) = n$, then

   $$\text{``}T_{p+1} H_k(X,\mathbb{Q}) = G_{p+1} H_k(X,\mathbb{Q}), \quad \forall k\text{''}$$

   is a birationally invariant statement for smooth projective varieties X with $\dim(X) \leq$

$n + 2$.

3.3 The proof of the Theorem 3.1.4

Proposition 3.3.1. *For any irreducible projective variety Y of dimension n, we have*

$$
\begin{cases}
L_{n-1}H_{2n}(X) \cong \mathbb{Z}, \\
L_{n-1}H_{2n-1}(X) \cong H_{2n-1}(X, \mathbb{Z}), \\
L_{n-1}H_{2n-2}(X) \to H_{2n-2}(X, \mathbb{Z}) \quad is \quad injective, \\
L_{n-1}H_k(X) = 0 \quad for \quad k > 2n.
\end{cases}
$$

Remark 3.3.1. *When Y is smooth projective, Friedlander have drawn a stronger conclusion, i.e., besides those in the proposition, $L_{n-1}H_{2n-2}(Y) \cong H_{n-1,n-1}(X, \mathbb{Z}) = NS(X)$.*

Proof. Set $S = Sing(Y)$, the set of singular points. Then S is the union of proper irreducible subvarieties. Set $S = (\cup_i S_i) \bigcup S'$, where $\dim(S_i) = n - 1$ and S' is the union of subvarieties with dimension $\leq n - 2$. Let $V = Y - S$ be the smooth open part of Y. According to Hironaka [Hi1], we can find $\tilde{Y}$ such that $\tilde{Y}$ is a smooth compactification of V. Let $D = \tilde{Y} - V$. D is a divisor on $\tilde{Y}$ with normal crossing. Denote by $i_0 : S \hookrightarrow Y$ and $i : D \hookrightarrow \tilde{Y}$ the inclusions of closed sets. Denote by $j_0 : V \hookrightarrow Y$ and $j : V \hookrightarrow \tilde{Y}$ the inclusions of open sets.

There are a few cases:

Case 1: $k \geq 2n$.

By the localization long exact sequence in Lawson homology

$$
\cdots \to L_{n-1}H_k(S) \to L_{n-1}H_k(Y) \to L_{n-1}H_k(V) \to L_{n-1}H_{k-1}(S) \to \cdots,
$$

we have

$$
L_{n-1}H_k(Y) \cong L_{n-1}H_k(V) \quad for \quad k \geq 2n
$$

since $L_{n-1}H_k(S) = 0$ for $k \geq 2n - 1$.

By the localization exact sequence in homology

$$
\cdots \to H_k(S) \to H_k(Y) \to H_k^{BM}(V) \to H_{k-1}(S) \to \cdots,
$$

we have

$$H_k(Y) \cong H_k^{BM}(V) \quad for \quad k \geq 2n$$

since $H_k(S) = 0$ for $k \geq 2n - 1$. Here $H_k^{BM}(V)$ is the Borel-Moore homology.

Similarly,

$$L_{n-1}H_k(\tilde{Y}) \cong L_{n-1}H_k(V) \quad for \quad k \geq 2n$$

and

$$H_k(\tilde{Y}) \cong H_k^{BM}(V) \quad for \quad k \geq 2n.$$

Since $\tilde{Y}$ is smooth, we have $L_{n-1}H_k(\tilde{Y}) \cong H_k(\tilde{Y})$ for $k \geq 2n$(cf. [F1]). This completes the proof for the case $k \geq 2n$.

Case 2: $k = 2n - 1$.

Applying Lemma 3.2.3 to the pair (Y, S) for $p = n - 1$, we have the commutative diagram of the long exact sequence

$$
\begin{array}{ccccccccc}
0 \to & L_{n-1}H_{2n-1}(Y) & \xrightarrow{j_0^*} & L_{n-1}H_{2n-1}(V) & \xrightarrow{(\delta_0)^*} & L_{n-1}H_{2n-2}(S) & \xrightarrow{(i_0)^*} & L_{n-1}H_{2n-2}(Y) & \to \\
& \downarrow \Phi_{n-1,2n-1} & & \downarrow \Phi_{n-1,2n-1} & & \downarrow \Phi_{n-1,2n-2} & & \downarrow \Phi_{n-1,2n-2} & \\
0 \to & H_{2n-1}(Y) & \xrightarrow{j_0^*} & H_{2n-1}^{BM}(V) & \xrightarrow{(\delta_0)^*} & H_{2n-2}(S) & \xrightarrow{(i_0)^*} & H_{2n-2}(Y) & \to
\end{array}
\tag{3.1}
$$

Similarly, applying Lemma 3.2.3 to the pair $(\tilde{Y}, D)$ for $p = n - 1$, we have the commutative diagram of the long exact sequence

$$
\begin{array}{ccccccccc}
0 \to & L_{n-1}H_{2n-1}(\tilde{Y}) & \xrightarrow{j^*} & L_{n-1}H_{2n-1}(V) & \xrightarrow{\delta_*} & L_{n-1}H_{2n-2}(D) & \xrightarrow{i_*} & L_{n-1}H_{2n-2}(\tilde{Y}) & \to & \cdots \\
& \downarrow \tilde{\Phi}_{n-1,2n-1} & & \downarrow \Phi_{n-1,2n-1} & & \downarrow \Phi_{n-1,2n-2} & & \downarrow \tilde{\Phi}_{n-1,2n-2} & \\
0 \to & H_{2n-1}(\tilde{Y}) & \xrightarrow{j^*} & H_{2n-1}^{BM}(V) & \xrightarrow{\delta_*} & H_{2n-2}(D) & \xrightarrow{i_*} & H_{2n-2}(\tilde{Y}) & \to & \cdots
\end{array}
\tag{3.2}
$$

Note that $\tilde{\Phi}_{n-1,2n-2} : L_{n-1}H_{2n-2}(\tilde{Y}) \to H_{2n-2}(\tilde{Y})$ is injective,

$$\tilde{\Phi}_{n-1,2n-1} : L_{n-1}H_{2n-1}(\tilde{Y}) \cong H_{2n-1}(\tilde{Y})$$

and $\tilde{\Phi}_{n-1,2n-2} : L_{n-1}H_{2n-2}(D) \cong H_{2n-2}(D) \cong \mathbb{Z}^m$, where m is the number of irreducible varieties of D. From (3.2) and the Five Lemma, we have the isomorphism

$$\Phi_{n-1,2n-1} : L_{n-1}H_{2n-1}(V) \cong H_{2n-1}^{BM}(V).$$
$$\tag{3.3}$$

From (3.1), (3.3) and the Five Lemma, we have the following isomorphism

$$\Phi_{n-1,2n-1} : L_{n-1}H_{2n-2}(Y) \cong H_{2n-2}(Y).$$

Case 3: $k = 2n - 2$.

Now the commutative diagram (3.1) is rewritten in the following way:

$$
\begin{array}{ccccccccc}
L_{n-1}H_{2n-1}(V) & \xrightarrow{(\delta_0)_*} & L_{n-1}H_{2n-2}(S) & \xrightarrow{(i_0)_*} & L_{n-1}H_{2n-2}(Y) & \xrightarrow{j_0^*} & L_{n-1}H_{2n-2}(V) & \to & 0 \\
\downarrow \Phi_{n-1,2n-1} & & \downarrow \Phi_{n-1,2n-2} & & \downarrow \Phi_{n-1,2n-2} & & \downarrow \Phi_{n-1,2n-2} & & \\
H_{2n-1}^{BM}(V) & \xrightarrow{(\delta_0)_*} & H_{2n-2}(S) & \xrightarrow{(i_0)_*} & H_{2n-2}(Y) & \xrightarrow{j_0^*} & H_{2n-2}^{BM}(V) & \to & 0
\end{array}
\tag{3.4}
$$

In the commutative diagram (3.2), we can show that the injective maps

$$
j^* : H_{2n-1}(\tilde{Y}) \to H_{2n-1}^{BM}(V)
\tag{3.5}
$$

and

$$
j^* : L_{n-1}H_{2n-1}(\tilde{Y}) \to L_{n-1}H_{2n-1}(V)
\tag{3.6}
$$

are actually isomorphisms. Hence the commutative diagram (3.2) reduces to the following diagram:

$$
\begin{array}{ccccccccc}
0 & \to & L_{n-1}H_{2n-2}(D) & \to & L_{n-1}H_{2n-2}(\tilde{Y}) & \to & L_{n-1}H_{2n-2}(V) & \to & 0 \\
& & \downarrow \Phi_{n-1,2n-2} & & \downarrow \tilde{\Phi}_{n-1,2n-2} & & \downarrow \Phi_{n-1,2n-2} & & \\
0 & \to & H_{2n-2}(D) & \to & H_{2n-2}(\tilde{Y}) & \to & H_{2n-2}^{BM}(V) & \to & 0
\end{array}
\tag{3.7}
$$

To see (3.5) are surjective, by the exactness of the rows in (3.2) we only need to show that the maps $i_* : H_{2n-2}(D) \to H_{2n-2}(\tilde{Y})$ are injective. Note that $\tilde{Y}$ is a compact Kähler manifold, and the homology class of an algebraic subvariety is nontrivial in the homology of the Kähler manifold. From these, we get the injectivity of i_*. The surjectivity of (3.6) follows from the same reason.

We need the following lemma.

Lemma 3.3.1. *The natural transformation* $\Phi_{n-1,2n-2} : L_{n-1}H_{2n-2}(V) \to H_{2n-2}^{BM}(V)$ *is injective.*

Proof. $a \in L_{n-1}H_{2n-2}(V)$ such that $\Phi_{n-1,2n-2}(a) = 0 \in H_{2n-2}^{BM}(V)$. Since the map $j^* : L_{n-1}H_{2n-2}(\tilde{Y}) \to L_{n-1}H_{2n-2}(V)$ is surjective, there exists an element $b \in L_{n-1}H_{2n-2}(\tilde{Y})$ such that $j^*(b) = a$. Set $\tilde{b} = \Phi_{n-1,2n-2}(b) \in H_{2n-2}(\tilde{Y})$. By the commutativity of the diagram, we have $j^*(\tilde{b}) = 0$ under the map $j^* : H_{2n-2}(\tilde{Y}) \to H_{2n-2}^{BM}(V)$. By the exactness of the bottom row in the commutative diagram (7), there exists an element $\tilde{c} \in H_{2n-2}(D)$ such that the image of $\tilde{c}$ under the map $i_* : H_{2n-2}(D) \to H_{2n-2}(\tilde{Y})$ is $\tilde{b}$. Now note that $\Phi_{n-1,2n-2} : L_{n-1}H_{2n-2}(D) \to H_{2n-2}(D)$ is an isomorphism, there exists an element $c \in L_{n-1}H_{2n-2}(D)$ such that $\Phi_{n-1,2n-2}(c) = \tilde{c}$. Hence $\Phi_{n-1,2n-2}(i_*(c) - b) = 0$. Note that $\Phi_{n-1,2n-2} : L_{n-1}H_{2n-2}(\tilde{Y}) \to H_{2n-2}(\tilde{Y})$ is injective since $\tilde{Y}$ is smooth and of dimension n (cf. [F1]). Hence we get $i_*(c) = b$, i.e., b is in the image of the map $i_* : L_{n-1}H_{2n-2}(D) \to$

$L_{n-1}H_{2n-2}(\tilde{Y})$. Therefore $a = 0$ by the exactness of the top row of the commutative diagram (3.7).

□

We need to show that $\Phi_{n-1,2n-2} : L_{n-1}H_{2n-2}(Y) \to H_{2n-2}(Y)$ is injective. For $a \in L_{n-1}H_{2n-2}(Y)$ such that $\Phi_{n-1,2n-2}(a) = 0 \in H_{2n-2}(Y)$. By the commutative diagram (3.4) and the Lemma 3.3.1, the image of a under $j_0^* : L_{n-1}H_{2n-2}(Y) \to L_{n-1}H_{2n-2}(V)$ is zero. Hence there exists an element $b \in L_{n-1}H_{2n-2}(S)$ such that the image of $(i_0)_* : L_{n-1}H_{2n-2}(S) \to L_{n-1}H_{2n-2}(Y)$ is a, i.e., $(i_0)_*(b) = a$. Set $\tilde{b} = \Phi_{n-1,2n-2}(b)$. Then the image of $\tilde{b}$ under the map $(i_0)_* : H_{2n-2}(S) \to H_{2n-2}(Y)$ is zero. By exactness of the bottom row in the commutative diagram (3.4), there exists an element $\tilde{c}$ such that its image under the map $H_{2n-1}^{BM}(V) \to H_{2n-2}(S)$ is $\tilde{b}$. By the result in **Case 2**, $\Phi_{n-1,2n-1} : L_{n-1}H_{2n-1}(V) \to H_{2n-1}^{BM}(V)$ is an isomorphism. Hence there exists an element $c \in L_{n-1}H_{2n-1}(V)$ such that $\Phi_{n-1,2n-1}(c) = \tilde{c}$. Now since $\Phi_{n-1,2n-2} : L_{n-1}H_{2n-2}(S) \to H_{2n-2}(S)$ is an isomorphism, the image of c under the map $L_{n-1}H_{2n-1}(V) \to L_{n-1}H_{2n-2}(S)$ is exactly b. Now the exactness of the top row of the commutative diagram (3.4) implies the vanishing of a.

The proof of the proposition is done.

□

By using this proposition, we will give a proof of Theorem 3.1.4.

Proof of Theorem 3.1.4:

For any smooth projective variety X, the injectivity of

$$T_p H_{2p+1}(X, \mathbb{Q}) \to G_p H_{2p+1}(X, \mathbb{Q})$$

has been proved in [[FM], §7]. We only need to show the surjectivity of $T_p H_{2p+1}(X, \mathbb{Q}) \to G_p H_{2p+1}(X, \mathbb{Q})$. For any subvariety $i : Y \subset X$, we denote by $V := X - Y$ the complementary of Y in X. We have the following commutative diagram of the long exact sequences (Lemma 3.2.3, or [Li1]):

$$
\begin{array}{ccccccccc}
\cdots \to & L_p H_{2p+1}(Y) & \to & L_p H_{2p+1}(X) & \to & L_p H_{2p+1}(V) & \to & L_p H_{2p}(Y) & \to & \cdots \\
& \downarrow \Phi_{p,2p+1} & & \downarrow \Phi_{p,2p+1} & & \downarrow \Phi_{p,2p+1} & & \downarrow \Phi_{p,2p} & & \\
\cdots \to & H_{2p+1}(Y) & \to & H_{2p+1}(X) & \to & H_{2p+1}^{BM}(V) & \to & H_{2p}(Y) & \to & \cdots
\end{array}
$$

Obviously, the above commutative diagram holds when tensored with $\mathbb{Q}$. In the following, we only consider the commutative diagrams with $\mathbb{Q}$-coefficient.

Now let $a \in G_p H_{2p+1}(X, \mathbb{Q})$, by definition, we can assume that a lies in the image of the map $i_* : H_{2p+1}(Y, \mathbb{Q}) \to H_{2p+1}(X, \mathbb{Q})$ for some subvariety $Y \subset X$ with dimension

$\dim Y = (2p + 1) - p = p + 1$. Hence there exists an element $b \in H_{2p+1}(Y, \mathbb{Q})$ such that $i_*(b) = a$. By the Proposition 3.3.1, we know that $\Phi_{p,2p+1} : L_p H_{2p+1}(Y) \otimes \mathbb{Q} \to H_{2p+1}(Y, \mathbb{Q})$ is an isomorphism. Therefore there exists an element $\tilde{b} \in L_p H_{2p+1}(Y) \otimes \mathbb{Q}$ such that $\Phi_{p,2p+1}(\tilde{b}) = b$. Set $\tilde{a} = i_*(\tilde{b})$. Then $\tilde{a}$ maps to a under the map $L_p H_{2p+1}(X) \otimes \mathbb{Q} \to H_{2p+1}(X, \mathbb{Q})$. By the definition of the topological filtration, $a \in T_p H_{2p+1}(X, \mathbb{Q})$. This completes the proof of surjectivity of $T_p H_{2p+1}(X, \mathbb{Q}) \to G_p H_{2p+1}(X, \mathbb{Q})$.

$\square$

Remark 3.3.2. *In the proof of the surjectivity in Theorem 3.1.4, the assumption of smoothness is not necessary, more precisely, for any irreducible projective variety X, the image of the natural transformation*

$$\Phi_{p,2p+1} : L_p H_{2p+1}(X, \mathbb{Q}) \to H_{2p+1}(X, \mathbb{Q})$$

contains $G_p H_{2p+1}(X, \mathbb{Q})$.

Remark 3.3.3. *Independently, M. Warker has recently also obtained this result ([Wa], Prop. 2.5]).*

Now we prove the corollaries 3.1.1-3.1.5.

The proof of Corollary 3.1.1: By Theorem 3.1.1 and 3.1.4, Dold-Thom Theorem and Proposition 3.3.1, we only need to show the cases that $p = 1, k \geq 5$. Now the following commutative diagram ([FM], Prop.6.3)

$$
\begin{array}{ccc}
L_2 H_k(X) \otimes \mathbb{Q} & \overset{s}{\to} & L_1 H_k(X) \otimes \mathbb{Q} \\
\downarrow \Phi_{2,k} & & \downarrow \Phi_{1,k} \\
H_k(X, \mathbb{Q}) & \cong & H_k(X, \mathbb{Q}).
\end{array}
$$

shows that if $L_2 H_k(X) \otimes \mathbb{Q} \to H_k(X, \mathbb{Q})$ is an surjective, then $L_1 H_k(X) \otimes \mathbb{Q} \to H_k(X, \mathbb{Q})$ must be surjective. Proposition 3.3.1 gives the needed surjectivity for $k \geq 5$ even if X is singular variety of dimension 3.

$\square$

The proof of Corollary 3.1.2: By Corollary 3.1.1, we only need to show that $T_1 H_4(X, \mathbb{Q}) = G_1 H_4(X, \mathbb{Q})$. By the assumption and Poincaré duality, $H_4(X, \mathbb{Q}) \cong H_2(X, \mathbb{Q}) \cong \mathbb{Q}$. Therefore, $G_1 H_4(X, \mathbb{Q}) = H_4(X, \mathbb{Q}) \cong \mathbb{Q}$ and again by the commutative diagram

$$
\begin{array}{ccc}
L_2 H_k(X) \otimes \mathbb{Q} & \overset{s}{\to} & L_1 H_k(X) \otimes \mathbb{Q} \\
\downarrow \Phi_{2,k} & & \downarrow \Phi_{1,k} \\
H_k(X, \mathbb{Q}) & \cong & H_k(X, \mathbb{Q}),
\end{array}
$$

we have the surjectivity of $L_1 H_4(X) \otimes \mathbb{Q} \to H_4(X, \mathbb{Q})$.

$\square$

The proof of Corollary 3.1.3: Suppose $X = S \times C$, where S is a smooth projective surface and C is a smooth projective curve. We only need to consider the surjectivity of $L_1 H_4(X) \otimes \mathbb{Q} \to H_4(X, \mathbb{Q})$ because of Corollary 3.1.1. Now the Künneth formula for the rational homology of $H_4(S \times C, \mathbb{Q})$ and Theorem 3.2.1 for S and C gives the surjectivity in this case.

$\square$

The proof of Corollary 3.1.4: This follows directly from Theorem 3.1.3.

$\square$

The proof of Corollary 3.1.5: By Theorem 3.1.4, we only need to show that

$$T_p H_k(X, \mathbb{Q}) = G_p H_k(X, \mathbb{Q})$$

for $k \geq 2p + 2$. By the definition of geometric filtration, an element $a \in G_p H_k(X, \mathbb{Q})$ comes from linear combinations of the images of elements $b_j \in H_k(Y_j, \mathbb{Q})$ for subvarieties Y_j of $\dim Y_j \leq k - p$ (equivalently, $\dim Y_j = k - p$). From the following commutative diagram

$$
\begin{array}{ccc}
L_p H_k(Y) \otimes \mathbb{Q} & \xrightarrow{i_* \otimes \mathbb{Q}} & L_p H_k(X) \otimes \mathbb{Q} \\
\downarrow{\scriptstyle \Phi_{p,k} \otimes \mathbb{Q}} & & \downarrow{\scriptstyle \Phi_{p,k} \otimes \mathbb{Q}} \\
H_k(Y, \mathbb{Q}) & \xrightarrow{i_* \otimes \mathbb{Q}} & H_k(X, \mathbb{Q}),
\end{array}
$$

it is enough to show that $\Phi_{p,k} : L_p H_k(Y) \to H_k(Y)$ is surjective for any irreducible subvariety $Y \subset X$ with $\dim(Y) = k - p$. By the Suslin conjecture, $\Phi_{p,k}$ is surjective for any smooth variety Y since $\dim(Y) = k - p$. Hence it is enough to show that $\Phi_{p,k}$ is also surjective for any singular irreducible variety Y under the assumption that the Sulin Conjecture for Lawson homology with coefficient $\mathbb{Z}$ holds.

We will show that the following lemma by induction:

Lemma 3.3.2. *If the Suslin Conjecture for Lawson homology with coefficient $\mathbb{Z}$ holds for every smooth projective variety, then the map $L_p H_k(Y) \to H_k^{BM}(Y)$ is an isomorphism for $k \geq m + p$ and a monomorphism for $k = m + p - 1$ for every (possibly singular) quasi-projective variety Y, where $m = \dim(Y)$.*

Proof. Suppose that Y is an irreducible quasi-projective variety with $\dim(Y) = m$, we now prove the lemma by induction on the dimension of Y. It is trivial if $m = \dim(Y) = 0$.

Let W be an irreducible quasi-projective variety with $\dim(W) = n < m$, by induction assumption, we have

$$\begin{cases} L_pH_{n+p-1}(W) \to H_{n+p-1}(W) & is \quad injective, \\ L_pH_{n+q}(W) \cong H_{n+q}(W) & for \quad q \geq p. \end{cases}$$

Denote by $\overline{Y}$ a projective closure of Y and $S = sing(\overline{Y})$ the singular point set of $\overline{Y}$. Set $U = \overline{Y} - S$. Let $\sigma : \widetilde{Y} \to \overline{Y}$ be a desingularization of $\overline{Y}$ and denote by $D := \widetilde{Y} - U$. The existence of a smooth $\widetilde{Y}$ is guaranteed by Hironaka [Hi1]. Then D is the union of irreducible varieties with dimension $\leq m - 1$.

By Lemma 3.3.2, we have the following commutative diagram

$$\begin{array}{ccccccccc} \cdots \to & L_pH_k(Z) & \to & L_pH_k(V) & \to & L_pH_k(U) & \to & L_pH_{k-1}(Z) & \to & \cdots \\ & \downarrow \Phi_{p,k} & & \downarrow \Phi_{p,k} & & \downarrow \Phi_{p,k} & & \downarrow \Phi_{p,k-1} & & \\ \cdots \to & H_k(Z) & \to & H_k(V) & \to & H_k^{BM}(U) & \to & L_pH_{k-1}(Z) & \to & \cdots, \end{array} \qquad (3.8)$$

where $U \subset V$ are quasi-projective varieties of $\dim(V) = \dim(U) = m$ and $Z = V - U$ is a closed subvariety of V.

Claim: By inductive assumption, the above commutative diagram (3.8) and the Five Lemma, we have the equivalence between

$$\begin{cases} L_pH_{m+p-1}(U) \to H_{m+p-1}(U) & is \quad injective, \\ L_pH_{m+q}(U) \cong H_{m+q}(U) & for \quad q \geq p. \end{cases}$$

and

$$\begin{cases} L_pH_{m+p-1}(V) \to H_{m+p-1}(V) & is \quad injective, \\ L_pH_{m+q}(V) \cong H_{m+q}(V) & for \quad q \geq p. \end{cases}$$

Using the Claim for finite times beginning from $V = \widetilde{Y}$, we have the result for any quasi-projective variety U and hence $\overline{Y}$ since S is the union of irreducible varieties of lower dimensions. Using the Claim once again we obtain the statement for Y since $\overline{Y} - Y$ is also the union of irreducible varieties of lower dimensions. The completes the proof of Lemma 3.3.2.

By Lemma 3.3.2, we have the Suslin Conjecture all for singular varieties if it holds for all smooth projective varieties. This completes the proof of Corollary 3.1.4. □

56

Chapter 4

A note on Lawson homology for smooth varieties with small Chow groups

4.1 Introduction

In this chapter, all projective varieties are defined over $\mathbb{C}$. Let X be a projective variety with dimension n. Let $\mathcal{Z}_p(X)$ be the space of algebraic p-cycles on X.

The **Lawson homology** $L_p H_k(X)$ of p-cycles is defined by

$$L_p H_k(X) = \begin{cases} \pi_{k-2p}(\mathcal{Z}_p(X)), & k \geq 2p; \\ 0, & k < 2p \end{cases}$$

where $\mathcal{Z}_p(X)$ is given a natural topology (cf. [F1], [L1]). For a general discussion of Lawson homology, see the survey paper [L2].

In [FM], Friedlander and Mazur showed that there are natural maps, called **cycle class maps**

$$\Phi_{p,k} : L_p H_k(X) \to H_k(X).$$

Definition 4.1.1. *Set*

$$L_p H_k(X)_{hom} := \ker\{\Phi_{p,k} : L_p H_k(X) \to H_k(X)\};$$

$$L_p H_k(X, \mathbb{Q})_{hom} := L_p H_*(X)_{hom} \otimes \mathbb{Q}.$$

C. Peters proved the following result by using the decomposition of the diagram for the smooth varieties with small Chow groups first shown by Bloch and Srinivas [BS] and generalized by Paranjape [Pa], Jannsen [J], Laterveer [Lat] and others:

Theorem 4.1.1. (Peters [Pe]) *Let X be a smooth projective variety for which rational and homological equivalence coincide for $p-$cycles in the range $0 \leq p \leq s$ (that is, in the terminology of [Lat], X has **small chow groups up to rank** s). Then $L_p H_*(X)_{hom} \otimes \mathbb{Q} = 0$ in the range $0 \leq p \leq s + 1$.*

By carefully checking the proof of Peters, we discover the symmetry of the decomposition of the diagonal $\Delta_X \subset X \times X$ and note that the proof works for p-cycles with $0 \leq n - p \leq s + 2$.

In this note, we will use the tools of Lawson homology and the methods and notations given in [Pe] (and the references therein) to show the following main result:

Theorem 4.1.2. *Let X be a smooth projective variety of dimension n for which rational and homological equivalence coincide for $p-$cycles in the range $0 \leq p \leq s$. Then $L_p H_*(X)_{hom} \otimes \mathbb{Q} = 0$ in the range $0 \leq n - p \leq s + 2$.*

For convenience, we introduce the following definition:

Definition 4.1.2. *A smooth projective variety X over $\mathbb{C}$ is called **rationally connected** if there is a rational curve through any 2 points of X. A necessary condition for Z to be rationally connected is that $\mathrm{Ch}_0(X) \cong \mathbb{Z}$.*

For equivalent descriptions of this definition, see the paper of Kollár, Miyaoka and Mori [KMM].

Corollary 4.1.1. *Let X be a smooth projective variety with $\dim(X) = 4$ and $\mathrm{Ch}_0(X) \cong \mathbb{Z}$. Then $L_p H_k(X)_{hom} \otimes \mathbb{Q} = 0$ for all p and k. In particular, all the smooth hypersurfaces in P^5 with degree less or equal than 5 have this property (cf. [Ro]).*

Remark 4.1.1. *It is shown by the author in [H1] that for any smooth projective rational variety X of $\dim(X) = 4$, $L_p H_k(X)_{hom} = 0$ for any p and k. Hence the nontriviality of $L_p H_k(X)_{hom}$ for some p,k for a rationally connected fourfold X would imply irrationality of X.*

Corollary 4.1.2. *Let X be a general cubic hypersurface of dimension less than or equal to 6, then $L_* H_*(X)_{hom} \otimes \mathbb{Q} = 0$.*

Remark 4.1.2. *Laterveer [Lat] showed that Griffiths groups are torsion for a general cubic hypersurface of dimension less than or equal to 6. For the general cubic sevenfold in P^8, Albano and Collino showed that $\mathrm{Griff}_3(X)$ (which is $\cong L_3 H_6(X)_{hom}$ by Friedlander in [F1]) is nontrivial even after tensoring with $\mathbb{Q}$.*

Remark 4.1.3. *This work was done in Spring of 2005 as part of my Ph. D. thesis. It was included in my research statement and put on my web page*

    http://www.math.sunysb.edu/~wenchuan/job/rs.pdf

in November 2005. I recently learned that M. Voineagu has independently obtained this result (cf. [Vn]).

4.2 The proof of the main result

The proof of the main theorem is based on: the Lemma 12 in [Pe], the decomposition of the diagonal given in [Pa], and the computation of Lawson homology of codimension 1 cycles for a smooth projective variety given by Friedlander [F1].

For convenience, we write the results we need as follows:

Theorem 4.2.1. (Friedlander [F1]) *Let X be any smooth projective variety of dimension n. Then we have the following isomorphisms*

$$\begin{cases} L_{n-1}H_{2n}(X) \cong \mathbb{Z}, \\ L_{n-1}H_{2n-1}(X) \cong H_{2n-1}(X,\mathbb{Z}), \\ L_{n-1}H_{2n-2}(X) \cong H_{n-1,n-1}(X,\mathbb{Z}) = NS(X) \\ L_{n-1}H_k(X) = 0 \quad for \quad k > 2n. \end{cases}$$

Remark 4.2.1. *From this theorem we have $L_{n-1}H_*(X)_{hom} = 0$ for any smooth projective variety X with $\dim(X) = n$.*

Now we need to review some definitions about the action of correspondences. Let X and Y be smooth projective varieties with $\dim(X) = n$. For $\alpha \in \mathcal{Z}_{n+d}(X \times Y)$, one puts

$$\alpha_*(u) = (p_2)_*[p_1^*(u) \cdot \alpha], \quad u \in \mathcal{Z}_p(X)$$

where $(p_2)_*$ is the proper push-forward, p_1^* is the flat pull back and the "." denotes the intersection product of cycles [[Pe], definition 10]. In this way, α_* gives a correspondence homomorphism

$$\alpha_* : \mathcal{Z}_p(X) \to \mathcal{Z}_{p+d}(Y).$$

This α_* induces a map (also denoted by α_*) on Lawson homology groups

$$\alpha_* : L_pH_k(X) \to L_{p+d}H_{k+2d}(Y)$$

which depends only on the class of α in the Chow group of $X \times Y$ modulo algebraic equivalence. For the details of the argument here, see [[Pe], section 1 C.]

The key Lemma we need was given by Peters as follows:

Proposition 4.2.1. *([Pe], Lemma 12) Assume that X and Y are smooth projective varieties and let $\alpha \subset X \times Y$ be an irreducible cycle of dimension $\dim(X) = n$, supported on $V \times W$, where, $V \subset X$ is a subvariety of dimension v and $W \subset Y$ a subvariety of dimension w. Let $\tilde{V}$, resp. $\tilde{W}$ be a resolution of singularities of V, resp. W and let $i : \tilde{V} \to X$ and $j : \tilde{W} \to Y$ be the corresponding morphisms. With $\tilde{\alpha} \subset \tilde{V} \times \tilde{W}$ the proper transform of α and p_1, resp. p_2 the projections from $X \times Y$ to the first. resp. the second factor, there is a commutative diagram*

$$
\begin{array}{ccc}
L_{p-n+v+w}H_{k+2(v+w-n)}(\tilde{V} \times \tilde{W}) & \xrightarrow{\tilde{\alpha}_*} & L_pH_k(\tilde{V} \times \tilde{W}) \\
\uparrow p_1^* & & \downarrow (p_2)_* \\
L_{p-n+v}H_{k+2(v-n)}(\tilde{V}) & & L_pH_k(\tilde{W}) \\
\uparrow i^* & & \downarrow j_* \\
L_pH_k(X) & \xrightarrow{\alpha_*} & L_pH_k(Y).
\end{array}
$$

Here i^* is induced by the Gysin homomorphism, p_1^* is the flat pull-back, and $(p_2)_*$ and j_* come from proper push forward. In particular, $\alpha_* = 0$ if $p < n - v$ or if $p > w$. Moreover, α_{n-v} acts trivially on $L_{n-v}H_*(X)_{hom}$, while α_w acts trivially on $L_wH_*(X)_{hom}$.

There is a corollary of this proposition given by Peters:

Corollary 4.2.1. *([Pe], Corollary 13) An irreducible cycle $\alpha \subset X \times X$ supported on a product variety $V \times W$ with $\dim V + \dim W = n = \dim(X)$ acts trivially on $L_*H_*(X)_{hom}$.*

Combining Friedlander's result (Theorem 4.2.1) and Peters' Lemma (Proposition 4.2.1), we have the following:

Corollary 4.2.2. *Under the assumptions of Proposition 4.2.1, we have that α_{w-1} acts trivially on $L_{w-1}H_*(X)_{hom}$.*

Now we want to recall some results about the decomposition of the diagonal given in [BS] and generalized by Paranjape [Pa] and Laterveer [Lat] with more general triviality hypotheses on the Chow group as stated in Theorem 4.1.1. Since the decomposition of diagonal is symmetric, we have the following version of the diagonal (cf. [V2], Theorem 10.29):

Theorem 4.2.2. *Let X be a smooth projective variety. Assume that for $p \leq s$, the maps*

$$cl : \mathrm{CH}_p(X) \otimes \mathbb{Q} \to H^{2n-2p}(X, \mathbb{Q})$$

are injective. Then there exists a decomposition

$$\Delta_X = \alpha^{(0)} + \cdots + \alpha^{(s)} + \beta \in \mathrm{CH}^n(X \times X) \otimes \mathbb{Q},$$

where $\alpha^{(p)}$ is supported in $V_p \times W_{n-p}$, $p = 0, \cdots, s$ with $\dim V_p = p$ and $\dim W_{n-p} = n-p$, and β is supported in $X \times W_{n-s-1}$.

Using the above theorem and Corollary 4.2.1, we deduce that the identity acts as β on the homologically zero part of the Lawson homology $L_* H_*(X)_{hom}$. Applying Proposition 4.2.1 and Corollary 4.2.2, we have the following main result:

Theorem 4.2.3. *Let X be a smooth projective variety such that the maps*

$$cl : \mathrm{CH}_p(X) \otimes \mathbb{Q} \to H^{2n-2p}(X, \mathbb{Q})$$

are injective for $p \leq s$. Then $L_{n-p}H_*(X)_{hom} \otimes \mathbb{Q} = 0$ for $p = 0, \cdots, s+1, s+2$.

As the application, we get Corollary 4.1.1 immediately.

Recall a result in [Pa] and [S], i.e., the general cubic hypersurface X of dimension greater than or equal to 5 has $\mathrm{Ch}_1(X) \cong \mathbb{Z}$ (Certainly $\mathrm{Ch}_0(X) \cong \mathbb{Z}$ by Roĭtman [Ro].) Hence we have the following

Corollary 4.2.3. *Let X be a general cubic hypersurface of dimension less than or equal 6, then $L_* H_*(X)_{hom} = 0$.* $\qquad\square$

Chapter 5

Infinitely generated Lawson homology groups on some rational projective varieties

5.1 Introduction

In this chapter we give examples of singular rational projective 4-dimensional varieties with infinitely generated Lawson homology groups even modulo torsion. This is totally different from the smooth case ([Pe], also [H1]), where it is known that all Lawson homology groups of rational fourfolds are finitely generated.

In this chapter we also give examples of singular rational projective 3-dimensional varieties with the same homeomorphism type but different Lawson homology groups.

For an algebraic variety X over $\mathbb{C}$, the **Lawson homology** $L_pH_k(X)$ of p-cycles is defined by

$$L_pH_k(X) := \pi_{k-2p}(\mathcal{Z}_p(X)), \quad k \geq 2p \geq 0$$

where $\mathcal{Z}_p(X)$ is provided with a natural topology. For general background, the reader is referred to the survey paper [L2].

Clemens showed that the **Griffiths group** of 1-cycles (which is defined to be the group of algebraic 1-cycles homologically equivalent to zero modulo l-cycles algebraically equivalent to zero) may be infinitely generated even modulo the torsion elements for general quintic hypersurfaces in $\mathbb{P}^4$ (cf. [Cl]). Friedlander showed that $L_1H_2(X)$ is exactly the algebraic 1-cycles modulo algebraic equivalence (cf. [F1]). Hence the Griffiths group of 1-cycles for X is a subgroup of $L_1H_2(X)$.

This leads to the following question:

(Q): Can one show that $L_p H_{2p+j}(X)$ is **not** finitely generated for some projective variety X where $j > 0$?

In this chapter we shall construct, for any given integers p and $j > 0$, examples of rational varieties X for which $L_p H_{2p+j}(X)$, as an abelian group, is infinitely generated. Thus, we answer affirmatively the above question :

Theorem 5.1.1. *There exists rational projective variety X with $\dim(X) = 4$ such that $L_1 H_3(X) \otimes \mathbb{Q}$ is **not** a finite dimensional $\mathbb{Q}$-vector space.*

By using the projective bundle theorem given by Friedlander and Gabber ([FG]), we have the following corollary:

Corollary 5.1.1. *For any $p \geq 1$, there exists projective algebraic variety X such that $L_p H_{2p+1}(X)$ is **not** a finitely generated abelian group.*

More generally, we have

Theorem 5.1.2. *For integers p and k, with $k \geq 0, p > 0$, we can find a projective variety Y, such that $L_p H_{2p+k}(Y)$ is infinitely generated.*

Remark 5.1.1. *The smoothness is essential here. Compare Theorem 5.1.1 with the following result proved by C. Peters.*

Theorem 5.1.3. *([Pe]) For any smooth projective variety X over $\mathbb{C}$ with $\mathrm{Ch}_0(X) \otimes \mathbb{Q} \cong \mathbb{Q}$, the natural map $\Phi : L_1 H_*(X) \otimes \mathbb{Q} \to H_*(X, \mathbb{Q})$ is injective. In particular, $L_1 H_*(X) \otimes \mathbb{Q}$ is a finite dimensional $\mathbb{Q}$-vector space.*

Any rational variety X (smooth or not) has the property that $\mathrm{Ch}_0(X) \otimes \mathbb{Q} \cong \mathbb{Q}$.

Applying the same construction to hypersurfaces in P^3, we obtain the following:

Theorem 5.1.4. *There exist two rational 3-dimensional projective varieties Y and Y' which are homeomorphic but for which the Lawson homology groups $L_1 H_3(Y, \mathbb{Q})$ and $L_1 H_3(Y', \mathbb{Q})$ are not isomorphic even up to torsion.*

Remark 5.1.2. *In fact, these varieties in Theorem 5.1.4 have exactly one isolated singular point.*

5.2 Lawson Homology

In this section we briefly review the definitions and results used in the next section. Let X be a projective variety of dimension m over $\mathbb{C}$. The group of p-cycles on X is the free abelian group $\mathcal{Z}_p(X)$ generated by irreducible p-dimensional subvarieties.

Definition 5.2.1. *The **Lawson homology** $L_pH_k(X)$ of p-cycles on X is defined by*

$$L_pH_k(X) := \pi_{k-2p}(\mathcal{Z}_p(X)), k \geq 2p \geq 0,$$

where $\mathcal{Z}_p(X)$ is provided with a natural, compactly generated topology (cf. [F1], [L1], [L2].

Definition 5.2.2. *The **Griffiths group** $\mathrm{Griff}_p(X)$ of p-cycles on X is defined by*

$$\mathrm{Griff}_p(X) := \mathcal{Z}_p(X)_{hom}/\mathcal{Z}_p(X)_{alg}$$

where $\mathcal{Z}_p(X)_{hom}$ denotes algebraic p-cycles homologous to zero and $\mathcal{Z}_p(X)_{alg}$ denotes algebraic p-cycles which are algebraically equivalent to zero.

Remark 5.2.1. *It was shown by Friedlander that $L_pH_{2p}(X) \cong \mathcal{Z}_p(X)/\mathcal{Z}_p(X)_{alg}$ (cf. [F1]). Hence the Griffiths group $\mathrm{Griff}_p(X)$ is a subgroup of the Lawson homology $L_pH_{2p}(X)$. Therefore, for any projective variety X (its homology groups are finitely generated), $\mathrm{Griff}_p(X)$ is infinitely generated if and only if $L_pH_{2p}(X)$ is.*

Remark 5.2.2. *For a quasi-projective variety U, $L_pH_k(U)$ is also well-defined and independent of the projective embedding (cf. [Li1], [L2]).*

Let $V \subset U$ be a Zariski open subset of a quasi-projective variety U. Set $Z = U - V$. Then we have

Theorem 5.2.1. *([Li1]) There is a long exact sequence for the pair (U, Z), i.e.,*

$$\cdots \to L_pH_k(Z) \to L_pH_k(U) \to L_pH_k(V) \to L_pH_{k-1}(Z) \to \cdots \qquad (5.1)$$

Remark 5.2.3. *For any quasi-projective variety U, $L_0H_k(U) \cong H_k^{BM}(U)$, where $H_k^{BM}(U)$ is the Borel-Moore homology. This follows from the Dold-Thom Theorem [DT].*

As a direct application of this long exact sequence, one has the following results [Li1]:

5.2.1 Let $U = \mathrm{P}^{n+1}$ and $V = \mathrm{P}^{n+1} - \mathrm{P}^n$. By the Complex Suspension Theorem [L1], we have, $L_pH_{2n}(\mathbb{C}^n) = \mathbb{Z}$; $L_pH_k(\mathbb{C}^n) = 0$ for any $k \neq 2n$ and $k \geq 2p \geq 0$.

5.2.2 Let $U = \mathbb{C}^n$ and $V_{n-1} \subset \mathbb{C}^n$ be a closed algebraic set. Set $V_n = \mathbb{C}^n - V_{n-1}$. Then we have

$$0 \to L_p H_{2n+1}(V_n) \to L_p H_{2n}(V_{n-1}) \to L_p H_{2n}(\mathbb{C}^n) \to L_p H_{2n}(V_n) \to L_p H_{2n-1}(V_{n-1}) \to 0$$

and

$$L_p H_{k+1}(V_n) \cong L_p H_k(V_{n-1}), \quad \text{k} \neq 2\text{n}, 2\text{n} + 1.$$

5.3   An Elementary Construction

Construction Let $X = (f(x_0, \cdots, x_{n+1}) = 0)$ be a general hypersurface in $\mathbb{P}^{n+1}$ with degree d, and let $V_n := X - X \cap \{\mathbb{P}^n = (x_0 = 0)\}$ be the affine part, i.e. $V_n \subset \mathbb{C}^{n+1}$. Define $V_{n+1} := \mathbb{C}^{n+1} - V_n$, then V_{n+1} can be viewed as an affine variety in $\mathbb{C}^{n+2}$ defined by $x_{n+2} \cdot f(1, x_1, \cdots, x_{n+1}) - 1 = 0$, where $V_n = (f(1, x_1, \cdots, x_{n+1}) = 0)$. Denoted by $\overline{V_{n+1}}$ the projective closure of V_{n+1} in $\mathbb{P}^{n+2}$ and set $Z_n = \overline{V_{n+1}} - V_{n+1}$.

We leave the study of this case where $n = 1$, and X is a smooth plane curve, as an exercise.

5.3.1   Application to the Case $n = 2$

In this subsection, I will show that there exist two rational projective 3-dimensional varieties with the same singular homology groups but different Lawson homology.

The following result proved by Friedlander will be used several times:

Theorem 5.3.1. *(Friedlander [F1]) Let X be any smooth projective variety of dimension n. Then we have the following isomorphisms*

$$\begin{cases} L_{n-1}H_{2n}(X) \cong \mathbb{Z}, \\ L_{n-1}H_{2n-1}(X) \cong H_{2n-1}(X, \mathbb{Z}), \\ L_{n-1}H_{2n-2}(X) \cong H_{n-1,n-1}(X, \mathbb{Z}) = NS(X) \\ L_{n-1}H_k(X) = 0 \quad for \quad k > 2n. \end{cases}$$

For a finitely generated abelian group G, we denote by $\mathrm{rk}(G)$ the rank of G.

Let $X \subset \mathrm{P}^3$ be a general surface with degree $d = 4$. Then $V_2 = X - X \cap \mathrm{P}^2$ and $C := X \cap \mathrm{P}^2$ is a smooth curve in P^2.

Lemma 5.3.1. $\mathrm{rk}(L_1 H_2(X)) \cong \mathrm{rk}(L_1 H_2(V_2)) + 1$; $\mathrm{rk}(L_1 H_3(V_2)) = 0$.

Proof. Applying Theorem 5.2.1 to the pair (X, C) and Theorem 5.3.1 for X, we get

$$0 \to L_1 H_3(V_2) \to L_1 H_2(C) \to L_1 H_2(X) \to L_1 H_2(V_2) \to 0.$$

Note that $L_1 H_2(C) \cong \mathbb{Z}$ and the map $L_1 H_2(C) \to L_1 H_2(X)$ is injective, and so we get $L_1 H_3(V_2) = 0$. Therefore, by the above long exact sequence, we have $\mathrm{rk}(L_1 H_2(X)) \cong \mathrm{rk}(L_1 H_2(V_2)) + 1$. $\qquad\square$

Lemma 5.3.2. $\mathrm{rk}(L_1 H_2(Z_2)) = 1$; $\mathrm{rk}(L_1 H_3(Z_2)) = 6$; $\mathrm{rk}(L_1 H_4(Z_2)) = 2$.

Proof. Note that $Z_2 = \overline{V_3} - V_3$ is defined by $(x_4 \cdot f(0, x_1, ..., x_3) = 0, x_0 = 0)$ in P^4. Let $C' = (x_4 = 0) \cap (f(0, x_1, \cdots, x_3) = 0)$ in the hyperplane $(x_0 = 0) \subset \mathrm{P}^3$. It is easy to see that $C' \cong C$. Then $Z_2 = \mathrm{P}^2 \cup \Sigma_p(C)$, where $\Sigma_p(C)$ means the joint of C and the point $p = [1 : 0 : \cdots : 0]$. By applying Theorem 5.2.1 to the pair $(Z_2, \Sigma C)$, we get

$$\cdots \to L_1 H_3(Z_2 - \Sigma C) \to L_1 H_2(\Sigma C) \to L_1 H_2(Z_2) \to L_1 H_2(Z_2 - \Sigma C) \to 0.$$

Note that

$$Z_2 - \Sigma C \cong \mathrm{P}^2 - C$$

and $L_1 H_3(\mathrm{P}^2 - C) = 0$. Therefore $\mathrm{rk}(L_1 H_2(Z_2)) = 1$. Moreover, since $L_1 H_4(\mathrm{P}^2 - C) = \mathbb{Z}$ and $L_1 H_4(\Sigma C) = \mathbb{Z}$ and $\mathrm{P}^2 \cap \Sigma C = C$ is a curve. The last statement follows. Recall that the Complex Suspension Theorem and Dold-Thom Theorem, we have $L_1 H_3(\Sigma C) \cong L_0 H_1(C) \cong H_1(C)$. By assumption, C is a plane curve of degree 4. The adjunction formula gives $\mathrm{rk}(H_1(C)) = 6$. The second statement follows. $\qquad\square$

Lemma 5.3.3. We have

$$\mathrm{rk}(L_1 H_2(\overline{V_3})) \leq 1;$$
$$\mathrm{rk}(L_1 H_3(\overline{V_3})) = \mathrm{rk}(L_1 H_2(X)) + \mathrm{rk}(L_1 H_2(\overline{V_3})) + 4.$$

Proof. Applying Theorem 5.2.1 to the pair $(\overline{V_3}, Z_2)$ with $p = 1$, we have

$$0 \to L_1 H_3(Z_2) \to L_1 H_3(\overline{V_3}) \to L_1 H_3(V_3) \to L_1 H_2(Z_2) \to L_1 H_2(\overline{V_3}) \to 0$$

67

since Lemma 5.3.1 gives $L_1 H_2(V_3) = 0$ and $L_1 H_4(V_3) \cong L_1 H_3(V_2) = 0$. Hence $\mathrm{rk}(L_1 H_2(\overline{V_3})) \leq 1$. Moreover, we have $\mathrm{rk}(L_1 H_3(Z_2)) - \mathrm{rk}(L_1 H_3(\overline{V_3})) + \mathrm{rk}(L_1 H_3(V_3)) - \mathrm{rk}(L_1 H_2(Z_2)) + \mathrm{rk}(L_1 H_2(\overline{V_3})) = 0$. By Lemma 5.3.2, we get

$$6 - \mathrm{rk}(L_1 H_3(\overline{V_3})) + (\mathrm{rk}(L_1 H_2(X)) - 1) - 1 + \mathrm{rk}(L_1 H_2(\overline{V_3})) = 0$$

$\square$

Lemma 5.3.4. $\mathrm{Sing}(\overline{V_3}) \cong \{X \cap (x_0 = 0)\} \cup \{p\} \cong C \cup \{p\}$.

Proof. It follows from a direct computation. By definition,

$$\mathrm{Sing}(\overline{V_3}) = \left\{ \begin{array}{l} F(x_0, x_1, x_2, x_3, x_4) = 0, \\ dF(x_0, x_1, x_2, x_3, x_4) = 0 \end{array} \right\}$$

$$= \left\{ \begin{array}{l} x_0^{d+1} - x_4 \cdot f(x_0, x_1, x_2, x_3) = 0, \\ (d+1)x_0 - x_4 \cdot \frac{\partial f}{\partial x_0} = 0, \\ -x_4 \cdot \frac{\partial f}{\partial x_1} = 0, \\ -x_4 \cdot \frac{\partial f}{\partial x_2} = 0, \\ -x_4 \cdot \frac{\partial f}{\partial x_3} = 0, \\ f(x_0, x_1, x_2, x_3) = 0 \end{array} \right\}$$

$$= \left\{ \begin{array}{l} x_0 = 0, \\ x_4 \cdot \frac{\partial f}{\partial x_0} = 0, \\ x_4 \cdot \frac{\partial f}{\partial x_1} = 0, \\ x_4 \cdot \frac{\partial f}{\partial x_2} = 0, \\ x_4 \cdot \frac{\partial f}{\partial x_3} = 0, \\ f(x_0, x_1, x_2, x_3) = 0 \end{array} \right\}$$

$$= \left\{ x_0 = x_3 = f(x_0, x_1, x_2) = 0 \right\} \cup \left\{ x_0 = \frac{\partial f}{\partial x_0} = \frac{\partial f}{\partial x_1} = \frac{\partial f}{\partial x_2} = \frac{\partial f}{\partial x_3} = 0 \right\}$$

$$\cong \left\{ x_0 = f(x_0, x_1, x_2) = 0 \right\} \cup \{p\} \cong C \cup \{p\}$$

since $C = (f = 0)$ is smooth by our assumption.

$\square$

Remark 5.3.1. *Note that p is an isolated singular point and the singularity $C = X \cap (x_0 = 0)$ is of A_n-type. We can resolve the singularity of this part by blowing up twice*

68

over the singularity, i.e., by blowing up over the singularity for the first time and then blowing up the singularity of the proper transform of the first blowup. We denote by $\widetilde{V_3}$ the proper transform of $\overline{V_3}$ with the exceptional divisor D_1 for the first blowup and $\widetilde{\widetilde{V_3}}$ the proper transform of $\widetilde{V_3}$ with the exceptional divisor D_2 for the second blowup. Both D_1 and D_2 are isomorphic to a fiber bundle over C with fibre the union of two P^1 intersecting at exactly one point. See the appendix for the computation of a concrete example.

Now $\widetilde{\widetilde{V_3}}$ has only one singular point, denote by q.

Lemma 5.3.5. *The singular point q in $\widetilde{\widetilde{V_3}}$ can be resolved by one blow up whose exceptional divisor is isomorphic to X.*

Proof. It follows from a trivial computation. $\quad\square$

We denote by W_3 the proper transform of the blow up in the above lemma. Note that W_3 is a **smooth** rational threefold. We have the following property on W_3:

Proposition 5.3.1. *For a smooth surface $X \subset \mathrm{P}^3$, the W_3 thus constructed is a smooth rational threefold with a fixed homeomorphic type, i.e., for two smooth surfaces X and X' in P^3, the corresponding smooth rational threefolds W_3 and W_3' are homeomorphic.*

Proof. Note that V_3 is a hypersurface in P^4. Let $(f_t(x_0, \cdots, x_4) = 0) \subset \mathrm{P}^4$ be a family of hypersurface such that $V_3 = (f_0 = 0)$ is transversal to the hypersurface $H = (x_0 = 0)$. Let Δ be a neighborhood of $t = 0$ such that $(f_t = 0)$ is transversal to H for all $t \in \Delta$. Let $W \subset \mathrm{P}^4 \times \Delta$ be the (analytic) variety defined by $F(x, t) := f_t(x) = 0$. Then we have the following incidence correspondence

$$
\begin{array}{ccc}
W & \subset & \mathrm{P}^4 \times \Delta \\
\downarrow \pi & & \downarrow \pi \\
\Delta & = & \Delta,
\end{array}
$$

with $\pi^{-1}(t) \cap W = W^t$.

By Remark 3.1 and Lemma 3.5, we get a smooth variety $\widetilde{W}$ by blowing up twice along 2-dimensional singularity of $Sing(W)$ and once for the remainder 1-dimensional singularity of $Sing(W)$. Denote by E the exceptional divisor of the last step. We claim that the map $\tilde{\pi} : \widetilde{W} \to \Delta$ is a smooth proper submersion. In fact, let v be a vector field of Δ and let $\tilde{v}$ be a lifting in $\Gamma(\widetilde{W}, T\widetilde{W})$ such that $\tilde{\pi}_*(\tilde{v}) = v$.

Denote by φ_t (resp. $\tilde{\varphi}_t$) the flow determined by v (resp. $\tilde{v}$). Then $\tilde{\varphi}_t : \widetilde{W}^0 \to \widetilde{W}^t$ gives the homeomorphism between two fiber of $\tilde{\pi}$ from Ehresmann's Theorem [V1]. This implies the result of the proposition. $\quad\square$

From this proposition, we have the following

Corollary 5.3.1. *For all smooth surfaces $X \subset \mathrm{P}^3$ of fixed degree, the $\widetilde{\widetilde{V_3}}$ thus constructed has a fixed homeomorphism type.*

Proof. In this proof of the proposition, we actually can choose $\tilde{v}$ such that 1)$\tilde{v}$ is tangent to W; 2) $\tilde{v}$ is tangent to the exceptional divisor E. Then the flow of $\tilde{v}$ gives the homeomorphism of any two fibers.

$\square$

We want to show that some Lawson homology group of $\widetilde{\widetilde{V_3}}$ may vary when the general X varies in P^3.

Theorem 5.3.2. *There exist two rational 3-dimensional projective varieties Y, Y' such that Y is homeomorphic to Y' but the Lawson homology group $L_1H_3(Y)$ is not isomorphic to $L_1H_3(Y')$ even up to torsion.*

Proof. If $X \subset \mathrm{P}^3$ is a general smooth quartic surface, then the Picard group $\mathrm{Pic}(X) \cong \mathbb{Z}$ by Noether-Lefschetz Theorem. For details, see e.g. Voisin [V2]. But it is well known that there are still many special smooth quartic surfaces X' in P^3 with $\mathrm{rk}(\mathrm{Pic}(X'))$ as big as 20. Note that by Theorem 5.3.1 and the Weak Lefschetz Theorem $L_1H_2(X) \cong \mathrm{Pic}(X)$ for any smooth surface X in P^3.

Now we choose smooth X with $L_1H_2(X) \cong \mathbb{Z}$ and X' with $L_1H_2(X') \cong \mathbb{Z}^{20}$. Set $Y := \widetilde{\widetilde{V_3}}$ and $Y' := \widetilde{\widetilde{V_3'}}$. Let W_3 (resp. W_3') be as in Proposition 5.3.1. From the proof of Lemma 2.1 in [H1], we have the commutative diagram

$$\cdots \to L_1H_3(E) \to L_1H_3(W_3) \to L_1H_3(W_3 - E) \to L_1H_2(E) \to \cdots$$
$$\downarrow \qquad\qquad \downarrow \qquad\qquad \downarrow\cong \qquad\qquad \downarrow$$
$$\cdots \to L_1H_3(q) \to L_1H_3(Y) \to L_1H_3(Y - q) \to L_1H_2(q) \to \cdots$$

By Lemma 3.5, we know $E \cong X$. By Theorem 5.3.1, we have $L_1H_3(X) \cong H_3(X)$. By the Lefschetz Hyperplane Theorem, we know X is simply connected. Since q is a point, we have $L_1H_3(Y) \cong L_1H_3(Y - q) \cong L_1H_3(W_3 - E)$ and $L_1H_2(Y) \cong L_1H_2(Y - q) \cong L_1H_2(W_3 - E)$.

The top row of the above commutative diagram turns into the long exact sequence

$$0 \to L_1H_3(W_3) \to L_1H_3(Y) \to L_1H_2(X) \to L_1H_2(W_3) \to L_1H_2(Y) \to 0$$

Therefore, we have

$$\mathrm{rk}L_1H_3(W_3) - \mathrm{rk}L_1H_3(Y) + \mathrm{rk}L_1H_2(X) - \mathrm{rk}L_1H_2(W_3) + \mathrm{rk}L_1H_2(Y) = 0$$

Since W_3 is a smooth rational threefold, we have $L_1H_3(W_3) \cong H_3(W_3)$, $L_1H_2(W_3) \cong H_2(W_3)$ ([[FHW], Prop. 6.16]) and by Proposition 5.3.1 $H_i(W_3) \cong H_i(W_3')$ for all i.

$$(*) \quad \mathrm{rk}L_1H_3(Y) \;=\; \mathrm{rk}H_3(W_3) + \mathrm{rk}L_1H_2(X) - \mathrm{rk}H_2(W_3) + \mathrm{rk}L_1H_2(Y)$$

By applying Theorem 5.2.1 to $(\widetilde{\widetilde{V_3}}, D_1)$, we get

$$\cdots \to L_1H_2(D_1) \to L_1H_2(\widetilde{\widetilde{V_3}}) \to L_1H_2(\widetilde{\widetilde{V_3}} - D_1) \to 0$$

Hence

$$
\begin{aligned}
\mathrm{rk}L_1H_2(\widetilde{\widetilde{V_3}}) \;&\leq\; \mathrm{rk}L_1H_2(D_1) + \mathrm{rk}L_1H_2(\widetilde{\widetilde{V_3}} - D_1) \\
&=\; \mathrm{rk}L_1H_2(D_1) + \mathrm{rk}L_1H_2(\widetilde{V_3} - q) \\
&=\; \mathrm{rk}L_1H_2(D_1) + \mathrm{rk}L_1H_2(\widetilde{V_3}) \\
&\leq\; \mathrm{rk}L_1H_2(D_1) + 1 \quad (\text{Lemma} \;\; 3.3)
\end{aligned}
$$

Similarly,

$$\mathrm{rk}L_1H_2(\widetilde{\widetilde{V_3}}) \;\leq\; \mathrm{rk}L_1H_2(D_2) + \mathrm{rk}L_1H_2(\widetilde{V_3})$$

Therefore,

$$\mathrm{rk}L_1H_2(\widetilde{\widetilde{V_3}}) \;\leq\; \mathrm{rk}L_1H_2(D_2) + \mathrm{rk}L_1H_2(D_1) + 1.$$

Since D_1 (also D_2) is isomorphic to a $\mathrm{P}^1 \cup \mathrm{P}^1$-bundle over a smooth curve C, it is easy to compute, by using Theorem 5.2.1 and the Projective Bundle Theorem [FG], that

$$\mathrm{rk}L_1H_2(D_1) \leq \mathrm{rk}L_1H_2(C) + 2 \cdot \mathrm{rk}L_0H_0(C) = 1 + 2 \times 1 = 3.$$

Therefore

$$\mathrm{rk}L_1H_2(\widetilde{\widetilde{V_3}}) \leq 3 + 3 + 1 = 7.$$

The same computation applies to $\widetilde{\widetilde{V_3'}}$ and we get

$$\mathrm{rk}L_1H_2(\widetilde{\widetilde{V_3'}}) \leq 3 + 3 + 1 = 7.$$

From this together with $(*)$, we have

$$
\begin{aligned}
\mathrm{rk}L_1H_3(\widetilde{\widetilde{V_3}}) \;&\leq\; \mathrm{rk}H_3(W_3) + \mathrm{rk}L_1H_2(X) - \mathrm{rk}H_2(W_3) + 7 \\
&=\; \mathrm{rk}H_3(W_3) - \mathrm{rk}H_2(W_3) + 8 \quad (\text{since} \;\; L_1H_2(X) \cong \mathbb{Z})
\end{aligned}
$$

On the other hand, we have

$$\operatorname{rk}L_1H_3(\widetilde{\widetilde{V_3'}}) \geq \begin{array}{l} \operatorname{rk}H_3(W_3') + \operatorname{rk}L_1H_2(X') - \operatorname{rk}H_2(W_3') \\ = \quad \operatorname{rk}H_3(W_3) + \operatorname{rk}L_1H_2(X') - \operatorname{rk}H_2(W_3) \\ = \quad \operatorname{rk}H_3(W_3) - \operatorname{rk}H_2(W_3) + 20 \quad (\text{since} \quad L_1H_2(X') \cong \mathbb{Z}^{20}) \end{array}$$

This shows that $L_1H_3(\widetilde{\widetilde{V_3}})$ is not isomorphic to $L_1H_3(\widetilde{\widetilde{V_3'}})$.

$\square$

5.3.2 Application to the Case $n = 3$

With this construction, if we choose $n = 3$ and $X \subset \mathrm{P}^4$ to be a general hypersurface of degree $d = 5$, then $V_3 = X - X \cap \mathrm{P}^3$ and $S := X \cap \mathrm{P}^3$ is a smooth surface in P^3.

The proof of Theorem 5.1.1: By applying Theorem 5.2.1 to the pair (X, S), we get

$$\cdots \rightarrow L_1H_3(V_3) \rightarrow L_1H_2(S) \rightarrow L_1H_2(X) \rightarrow L_1H_2(V_3) \rightarrow 0. \quad (5.2)$$

The above long exact sequence (5.2) remains exact after being tensored with $\mathbb{Q}$. Note that $L_1H_2(X) \otimes \mathbb{Q} \supset \operatorname{Griff}_1(X) \otimes \mathbb{Q}$ is an infinite dimensional $\mathbb{Q}$-vector space by [Cl]. Recall that $L_1H_2(S)$ is finitely generated since $\dim S = 2$ (cf. [F1]). Hence $L_1H_2(V_3) \otimes \mathbb{Q}$ is an infinite dimensional $\mathbb{Q}$-vector space. By (5.2.2), we have $L_1H_3(V_4) \otimes \mathbb{Q} \cong L_1H_2(V_3) \otimes \mathbb{Q}$ is an infinite dimensional $\mathbb{Q}$-vector space.

Note that $Z_3 = \overline{V_4} - V_4$ is defined by $(x_5 \cdot f(0, x_1, ..., x_4) = 0, x_0 = 0)$ in P^5. Let $S' = (x_5 = 0) \cap (f(0, x_1, ..., x_4) = 0)$ in the hyperplane $(x_0 = 0) \subset \mathrm{P}^5$. It is easy to see that $S' \cong S$. Then $Z_3 = \mathrm{P}^3 \cup \Sigma_p(S)$, where $\Sigma_p(S)$ means the joint of S and the point $p = [1 : 0 : \cdots : 0]$. By applying Theorem 5.2.1 to the pair $(Z_3, \Sigma S)$, we get

$$\cdots \rightarrow L_1H_3(Z_3 - \Sigma S) \rightarrow L_1H_2(\Sigma S) \rightarrow L_1H_2(Z_3) \rightarrow L_1H_2(Z_3 - \Sigma S) \rightarrow 0. \quad (5.3)$$

Note that $Z_3 - \Sigma S \cong \mathrm{P}^3 - S$. Therefore $L_1H_2(Z_3) \otimes \mathbb{Q}$ is of finite dimensional since both $L_1H_2(\sum S) \otimes \mathbb{Q} \cong L_0H_0(S, \mathbb{Q}) \cong \mathbb{Q}$ ([L1]) and $L_1H_2(\mathrm{P}^3 - S) \otimes \mathbb{Q}$ are. By the same type argument, we have $L_1H_3(Z_3) \otimes \mathbb{Q}$ is of finite dimensional since both $L_1H_3(\sum S) \otimes \mathbb{Q} \cong L_0H_1(S, \mathbb{Q}) = 0$(note that S is simply connected) and $L_1H_3(\mathrm{P}^3 - S) \otimes \mathbb{Q}$ are.

By applying Theorem 5.2.1 to the pair $(\overline{V_4}, Z_3)$, we have the following long exact

72

sequence

$$\cdots \to L_1H_3(Z_3) \to L_1H_3(\overline{V_4}) \to L_1H_3(V_4) \to L_1H_2(Z_3) \to \cdots \qquad (5.4)$$

From (5.4), the infinite dimensionality of $L_1H_3(V_3) \otimes \mathbb{Q}$, the finite dimensionality of $L_1H_2(Z_3) \otimes \mathbb{Q}$ and $L_1H_3(Z_3) \otimes \mathbb{Q}$, we obtain that $L_1H_3(\overline{V_4}) \otimes \mathbb{Q}$ is an infinitely dimensional $\mathbb{Q}$-vector space. This completes the proof of Theorem 5.1.1. $\qquad\square$

We can continue the procedure. Set $V_5 := \mathbb{C}^5 - V_5$, then V_5 can be viewed as an affine variety in $\mathbb{C}^6$ defined by $x_6 \cdot (x_5 \cdot f(1, x_1, \cdots, x_4) - 1) - 1 = 0$. Set $Z_4 = \overline{V_5} - V_5$, and so on. It can be shown in the same way that $L_1H_3(Z_4)$ is finitely generated by using Theorem 5.2.1 and Lawson's Complex Suspension Theorem. Note that $L_1H_4(V_5) \cong L_1H_3(V_4)$ is infinitely generated by 2.2.

By applying Theorem 5.2.1 to the pair $(\overline{V_5}, Z_4)$, we get the long exact sequence

$$\to L_1H_4(Z_4) \to L_1H_4(\overline{V_5}) \to L_1H_4(V_5) \to L_1H_3(Z_4) \to \cdots$$

From these we obtain that $L_1H_4(\overline{V_5})$ is infinitely generated.

Proposition 5.3.2. *In this construction, $L_1H_k(\overline{V_{k+1}})$ is **not** finitely generated for $k \geq 3$.*
$\qquad\square$

From the Complex Suspension Theorem [L1], we have

$$L_{p+1}H_{2p+k}(\Sigma^p \overline{V_{k+1}}) \cong L_1H_k(\overline{V_{k+1}}).$$

Therefore we get:

Theorem 5.3.3. *For integers p and k, with $k > 0, p > 0$, we can find a rational projective variety Y, such that $L_pH_{2p+k}(Y)$ is infinitely generated.*
$\qquad\square$

Remark 5.3.2. *If $k = 0$ and $p > 0$, there also exists projective varieties Y such that $L_pH_{2p}(Y)$ is infinitely generated. This follows from the Projective Bundle Theorem [FG] and a result of Clemens [Cl].*

Remark 5.3.3. *All the Y thus constructed above are singular projective varieties. Can one find some smooth projective variety such that the answer to the question (Q) is positive? Yes, we can. The author has constructed examples of smooth projective varieties such (Q) is true (cf. [H6]).*

Remark 5.3.4. *Note that all $\overline{V_{k+1}}$ are singular rational projective varieties. For smooth rational projective varieties Y, $L_1H_*(Y) \otimes \mathbb{Q}$ are finite dimensional $\mathbb{Q}$-vector spaces [Pe]. The author showed $L_1H_*(Y)$ are finitely generated abelian groups [H1].*

73

5.4 Appendix

Let $f(x_0, \cdots, x_4)$ be a general homogenous polynomial of degree 5 and X be a hypersurface of degree 6 in P^5 given by $F(x_0, \cdots, x_5) := x_5 f(x_0, \cdots, x_4) - x_0^6 = 0$. It is easy to see from the proof of Lemma 5.3.4 that the singular points set of X is the union of a smooth 2-dimensional variety Y given by $x_0 = x_5 = f(x_0, x_1, \cdots, x_4) = 0$ and an isolated point defined by $\{x_0 = \frac{\partial f}{\partial x_0} = \frac{\partial f}{\partial x_1} = \frac{\partial f}{\partial x_2} = \cdots = \frac{\partial f}{\partial x_5} = 0\}$.

Let $\sigma : \widetilde{\mathrm{P}^5_Y} \to \mathrm{P}^5$ be the blow up of P^5 along the surface Y and $\tilde{X}_Y$ be the proper transform in the blow up $\widetilde{\mathrm{P}^5_Y}$. Denoted by $E = \mathrm{P}(N_{Y/\mathrm{P}^5})$ the exceptional divisor of the blow-up. Then $D = E \cap \tilde{X}_Y \subset \mathrm{P}(N_{Y/\mathrm{P}^5})$ corresponds to the image of the tangent cones $T_p X \subset T_p(\mathrm{P}^5)$ in $\mathrm{P}(N_{Y/\mathrm{P}^5})$ at points $p \in Y$.

Now

$$T_p X = \left\{ \sum_{i_0 + \cdots i_5 = 2} \frac{\partial^2 F}{\partial x_0^{i_0} \cdots \partial x_5^{i_5}} x_0^{i_0} \cdots x_5^{i_5} = 0 \right\}$$

a degree 2 polynomial in P^5. Directly computation shows that

$$T_p X = \left\{ \frac{\partial f}{\partial x_0}(p) x_0 x_5 + \cdots + \frac{\partial f}{\partial x_4}(p) x_4 x_5) = 0 \right\}$$

$$= (x_5 = 0) \cup \left\{ \frac{\partial f}{\partial x_0}(p) x_0 + \cdots + \frac{\partial f}{\partial x_4}(p) x_4) = 0 \right\}.$$

Hence $D = \tilde{X}_Y \cap E$ is a fiber bundle over Y with singular conics as fibers. Clearly, $\tilde{X}_Y$ is smooth away from D. Since $D \subset E$ is a 3-dimensional variety with singular points set $S \cong Y$, we can show that it is the only singularity on $\tilde{X}_Y$:

Proposition 5.4.1. *The proper transform $\tilde{X}_Y$ is a 4-dimensional variety in $\mathrm{P}(N_{Y/\mathrm{P}^5})$ with singularity $S \cong Y \cup \{q\}$, where q is an isolated singular point.*

Proof. From the proof of Lemma 5.3.4, we see that the singular points S of X consist of two components. One is a smooth surface and the other is an isolated point q.

Since f is nonsingular on $Y = \{x_0 = f(0, x_1, \cdots, x_4) = 0\}$, we have $df \neq 0$ on Y. Let us restrict ourselves to a neighborhood of a point p in Y. There, we can take the neighborhood of p as the affine space $\mathbb{C}^5$ with p the origin. Hence we can choose $y = f$ as a coordinate in the neighborhood of each point on Y since it is smooth. Locally, Y is defined by $x_0 = x_5 = y = 0$ in $\mathbb{C}^5$. We denote it by Y_0. For convenience, we denote x_0 by x, x_5 by z. The blow up $\widetilde{(\mathbb{C}^5)}_{Y_0}$ of $\mathbb{C}^5$ along Y_0 is defined by the system of equations

$$\begin{cases} xv = uy, \\ xw = uz, \\ yw = zv. \end{cases}$$

74

in $\mathbb{C}^5 \times P^2$, where $[u : v : w]$ is the homogenous coordinates on P^2. Let $\sigma : \widetilde{(\mathbb{C}^5)}_{Y_0} \to \mathbb{C}^5$ be the map of this blowup. Then the inverse image of X is given by the following equations:

$$\begin{cases} x^6 - yz = 0, \\ xv = uy, \\ xw = uz, \\ yw = zv. \end{cases}$$

The above equations define two divisors on $\widetilde{(\mathbb{C}^5)}_{Y_0}$. One of them is the exceptional divisor E_0, the intersection of E with $\widetilde{(\mathbb{C}^5)}_{Y_0}$ and the other is exactly the proper transform $\widetilde{X}_{Y_0}$ of X_0 in $\widetilde{(\mathbb{C}^5)}_{Y_0}$, where X_0 is the part of X in $\mathbb{C}^5$.

We want to show that $\widetilde{X}_{Y_0}$ is smooth away from Y_0. Now it is clear. The blow up $\widetilde{(\mathbb{C}^5)}_{Y_0}$ is covered by 3 open charts: $(u \neq 0)$, $(v \neq 0)$ and $(w \neq 0)$.

On the chart $(u \neq 0)$, we can set $u = 1$. The equations for the inverse image of X_0 under σ are given by

$$\begin{cases} x^6 - yz = 0, \\ xv = y, \\ xw = z, \\ yw = zv. \end{cases}$$

The equations $xv = y$ and $xw = z$ imply $yw = zv$. Replacing y and z by xv and xw, respectively, we can factor x^2 in the first equation $x^6 - (xv)(xw) = 0$. Hence the proper transform $\widetilde{X}_{Y_0}$ are given by

$$\begin{cases} x^4 - vw = 0, \\ xv = y, \\ xw = z. \end{cases}$$

and the exceptional divisor is given by

$$\begin{cases} x^2 = 0, \\ xv = y, \\ xw = z. \end{cases}$$

i.e., $x = y = z = v = w = 0$, which is isomorphic to Y_0.

It is easy to show that on the charts $(v \neq 0)$ and $(w \neq 0)$, the proper transform $\widetilde{X}_{Y_0}$ is smooth everywhere. This completes the proof of the proposition.

$\square$

Remark 5.4.1. *In fact, the 2-dimensional singularity of X is of the A_n-type. It can be resolved by blowing up one more time. The isolated singularity q can be resolved by one blowup.*

Chapter 6

Generalized Abel-Jacobi map on Lawson homology

6.1 Introduction

In this chapter, all varieties are defined over $\mathbb{C}$. Let X be a smooth projective variety with dimension n. Recall that the **Hodge filtration**

$$\cdots \subseteq F^i H^k(X, \mathbb{C}) \subseteq F^{i-1} H^k(X, \mathbb{C}) \subseteq \cdots \subseteq F^0 H^k(X, \mathbb{C}) = H^k(X, \mathbb{C})$$

is defined by

$$F^q H^k(X, \mathbb{C}) := \bigoplus_{i \geq q} H^{i,k-i}(X).$$

Note that $F^q H^k(X, \mathbb{C})$ vanishes if $q > k$.

In [G], Griffiths generalized the Jacobian varieties and the Abel-Jacobi map on smooth algebraic curves to higher dimensional smooth projective varieties.

Definition 6.1.1. *The **q-th intermediate Griffiths Jacobian** of a smooth projective variety X is defined by*

$$
\begin{aligned}
J^q(X) : \; &= \; H^{2q-1}(X, \mathbb{C}) / \{ F^q H^{2q-1}(X, \mathbb{C}) + H^{2q-1}(X, \mathbb{Z}) \} \\
&\cong \; F^{n-q+1} H^{2n-2q+1}(X, \mathbb{C})^* / H^{2q-1}(X, \mathbb{Z})^*.
\end{aligned}
$$

Let $\mathcal{Z}_p(X)$ be the space of algebraic p-cycles on X. Set $\mathcal{Z}^{n-p}(X) \equiv \mathcal{Z}_p(X)$. There is a natural map

$$cl_q : \mathcal{Z}^q(X) \to H^{2q}(X, \mathbb{Z})$$

called **the cycle class map**. Set

$$\mathcal{Z}_{n-q}(X)_{hom} := \mathcal{Z}^q(X)_{hom} := \ker cl_q.$$

Definition 6.1.2. *The **Abel-Jacobi map***

$$\Phi^q : \mathcal{Z}^q(X)_{hom} \to J^q(X)$$

sends $\varphi \in \mathcal{Z}^q(X)_{hom}$ to Φ^q_φ, where Φ^q_φ is defined by

$$\Phi^q_\varphi(\omega) := \int_U \omega, \quad \omega \in F^{n-q+1} H^{2n-2q+1}(X, \mathbb{C}).$$

Here $\varphi = \partial U$ and U is an integral current of dimension $2n - 2q + 1$.

Now let

$$J^{2q-1}(X)_{alg} \subseteq J^{2q-1}(X)$$

be the largest complex subtorus of $J^{2q-1}(X)$ whose tangent space is contained in $H^{q-1,q}(X)$. It has been proved that $\Phi^q(\mathcal{Z}^q(X)_{alg})$ is a subtorus of $J^{2q-1}(X)$ contained in $J^{2q-1}(X)_{alg}$ (cf. [V1], Corollary 12.19), where $\mathcal{Z}^q(X)_{alg} \subseteq \mathcal{Z}^q(X)$ are the subset of codimension q-cycles which are algebraically equivalent to zero.

The **Griffiths group** of codimension q-cycles is defined to

$$\text{Griff}^q(X) := \mathcal{Z}^q(X)_{hom}/\mathcal{Z}^q(X)_{alg}$$

Therefore we can define the transcendental part of the Abel-Jacobi map

$$\Phi^q_{tr} : \text{Griff}^q(X) \to J^q(X)_{tr} := J^{2q-1}(X)/J^{2q-1}(X)_{alg} \tag{6.1}$$

as the factorization of Φ^q.

By using this, Griffiths showed the following:

Theorem 6.1.1. ([G]) *Let $X \subset P^4$ be a general quintic 3-fold, the Griffiths group* $\mathrm{Griff}^2(X)$ *is nontrivial, even modulo torsion.*

Remark 6.1.1. *Clemens has obtained further results: Under the same assumption as in Theorem 6.1.1, $\mathrm{Griff}^2(X) \otimes \mathbb{Q}$ is an infinitely generated $\mathbb{Q}$-vector space [Cl].*

In this chapter, the Griffiths' Abel-Jacobi map is generalized to the spaces of the homologically trivial part of Lawson homology groups.

Definition 6.1.3. *The **Lawson homology** $L_p H_k(X)$ of p-cycles is defined by*

$$L_p H_k(X) := \pi_{k-2p}(\mathcal{Z}_p(X)) \quad for \quad k \geq 2p \geq 0,$$

where $\mathcal{Z}_p(X)$ is provided with a natural topology (cf. [F1], [L1]). For general background, the reader is referred to [L2].

In [FM], Friedlander and Mazur showed that there are natural maps, called **cycle class maps**

$$\Phi_{p,k} : L_p H_k(X) \to H_k(X).$$

Define

$$L_p H_k(X)_{hom} := \ker\{\Phi_{p,k} : L_p H_k(X) \to H_k(X)\}.$$

The domain of Abel-Jacobi map can be reduced to Griffiths groups as in (6.1). Similarly, our generalized Abel-Jacobi map is defined on homologically trivial part of Lawson homology groups. As an application, we show that the non-triviality of certain Lawson homology group, even modulo the usual homology.

The main result in this chapter is the following:

Theorem 6.1.2. *Let X be a smooth projective variety. There is a well-defined map*

$$\Phi : L_p H_{2p+k}(X)_{hom} \longrightarrow \left\{ \bigoplus_{r > k+1, r+s=k+1} H^{p+r,p+s}(X) \right\}^* \Big/ H_{2p+k+1}(X, \mathbb{Z})$$

which generalizes Griffiths' Abel-Jacobi map defined in [G]. Moreover, for any $p > 0$ and $k \geq 0$, we find examples of projective manifolds X for which the image of the map on $L_p H_{k+2p}(X)_{hom}$ is infinitely generated.

As the application of the main result together Clemens' Theorem (Remark 6.1.1), we obtain

Theorem 6.1.3. *For any $k \geq 0$, there exist a projective manifold X of dimension $k + 3$ such that $L_1 H_{k+2}(X)_{hom} \otimes \mathbb{Q}$ is nontrivial, in fact, infinite dimensional over $\mathbb{Q}$.*

Using the Projective Bundle Theorem proved by Friedlander and Gabber in [FG], we have the following result:

Theorem 6.1.4. *For any $p > 0$ and $k \geq 0$, there exist a smooth projective variety X such that $L_p H_{k+2p}(X)_{hom} \otimes \mathbb{Q}$ is an infinite dimensional vector space over $\mathbb{Q}$.*

In section 6.2, we will review the minimal background materials about Lawson homology and point out its relation to Griffiths groups. In section 6.3, we give the definition the generalized Abel-Jacobi map. In section 6.4, the non-triviality of the generalized Abel-Jacobi map is proved by using Griffiths and Clemens' results through examples. The construction in our examples also shows this generalized Abel-Jacobi map really generalizes Griffiths' result in [G].

6.2 Lawson homology

Let X be a projective variety of dimension n. Denote by $\mathcal{C}_p(X)$ be the space of effective algebraic p-cycles on X and $\mathcal{Z}_p(X)$ be the space of algebraic p-cycles on X. There is natural, compactly generated topology on $\mathcal{C}_p(X)$ (resp. $\mathcal{Z}_p(X)$) and therefore $\mathcal{C}_p(X)$ (resp. $\mathcal{Z}_p(X)$) carries a structure of an abelian topological group.

The **Lawson homology** $L_p H_k(X)$ of p-cycles is defined by

$$L_p H_k(X) := \pi_{k-2p}(\mathcal{Z}_p(X)) \quad for \quad k \geq 2p \geq 0.$$

It has been proved by Friedlander in [F1] that

$$L_p H_k(X) \cong \varinjlim \pi_k(\mathcal{C}_p(X)_\alpha)$$

for all $k > 0$, where the limit is taken over the connected components of $\mathcal{C}_p(X)$ with respect to the action of $\pi_0(\mathcal{Z}_p(X))$. For a detailed discussion of this construction and its properties we refer the reader to [FM], §2 and [FL1],§1.

In [FM], Friedlander and Mazur showed that there are natural maps, called **cycle class maps**

$$\Phi_{p,k} : L_p H_k(X) \to H_k(X)$$

where $H_k(X)$ is the singular homology with the integral coefficient.

Define

$$L_p H_k(X)_{hom} := \ker\{\Phi_{p,k} : L_p H_k(X) \to H_k(X)\}.$$

80

It was proved by Friedlander [F1] that $L_p H_{2p}(X) \cong \mathcal{Z}_p(X)/\mathcal{Z}_p(X)_{alg}$. Therefore we have

$$L_p H_{2p}(X)_{hom} \cong \mathrm{Griff}_p(X), \qquad (6.2)$$

where $\mathrm{Griff}_p(X) := \mathrm{Griff}^{n-p}(X)$.

For general background on Lawson homology, the reader is referred to [L2].

6.3 The definition of generalized Abel-Jacobi map on

$$L_p H_{2p+k}(X)_{hom}$$

In this section, X denotes a smooth projective algebraic manifold with dimension n. Note that $\mathcal{Z}_p(X)$ is an abelian topological group with an identity element, the "null" p-cycle.

For $[\varphi] \in L_p H_{2p+k}(X)$, we can construct an integral $(2p+k)$-cycle c in X.

To see how to construct c from $[\varphi]$ for the case that $p = 0$, the reader is referred to [FL3]. The construction that φ is morphism from a projective variety to $\mathcal{Z}_p(X)$.

We will use this construction several times in the following. We briefly review the construction here.

A class

$$[\varphi] \in L_p H_{2p+k}(X) = \varinjlim \pi_k(\mathcal{C}_j(X))$$

is represented by a map

$$\varphi : S^k \to \mathcal{C}_p(X).$$

(For $k = 0$, $[\varphi]$ is represented by a difference of such maps.)

We may assume φ to be piecewise linear (PL for short) with respect to a triangulation of $\mathcal{C}_p(X) \supset \Gamma_1 \supset \Gamma_2 \supset \cdots$ respecting the smooth stratified structure ([Hi2]). Here Γ_i is a subcomplex for every $i > 0$.

Let φ be as above and fix $s_0 \in S^k$ and $x_0 \in \mathrm{Supp}\ (\varphi(s_0)) \subset X$. There exist affine coordinates $(z_1, \cdots z_p, \zeta_1, \cdots, \zeta_{n-p})$ on X with $x_0 = 0$ such that the projection $pr_1(z, \zeta) = z$, when restricted to $U \times U' = \{(z, w) : |z| < 1\ and\ |w| < 1\}$, gives a proper (finite) map $pr_1 : \mathrm{Supp}\ (\varphi(s_0)) \cap (U \times U') \to U$. Slicing this cycle $\varphi(s_0)|_{U \times U'}$ by this projection gives a PL map $\sigma : U \to SP^d(U')$ (with respect to a triangulation of $SP^d(U')$) for some d. Furthermore, given any such a map, we can construct a cycle in $U \times U'$. (cf. [FL3].) Choose a finite number of such product neighborhood $U_\alpha \times U'_\alpha$, $\alpha = 1, \cdots, K$, so that the union of $U_\alpha \times U'_\alpha(\frac{1}{2})$ covers $\mathrm{Supp}\ (\varphi(s_0))$. After shrinking each U'_α slightly, we can find a neighborhood $\mathcal{N}$ of s_0 in S^k such that $pr : \mathrm{Supp}\ (\varphi(s)) \cap (U_\alpha \times U'_\alpha) \to U_\alpha$ for all $s \in \mathcal{N}$

and for all α. Then φ is PL in $\mathcal{N}$ if and only if $\sigma : \mathcal{N} \times U_\alpha \to SP^d(U_\alpha')$ is PL for all α. One defines the cycle $c(\varphi)$ in each neighborhood $\mathcal{N} \times U_\alpha \times U_\alpha'$ by graphing this extended σ. From the construction, the cycle $c(\varphi)$ depends only on the PL map φ. (The argument here is from [[FL1], page 370-371].)

Lemma 6.3.1. *The homology class $c_\varphi := (pr_2)_*(c(\varphi))$ is independent of the choice of PL map $\varphi : S^k \to \mathcal{C}_p(X)$ in $[\varphi]$, where $pr_2 : S^k \times X \to X$ is the projection onto the second factor.*

Proof. Suppose that $\varphi' : S^k \to \mathcal{C}_p(X)$ is another PL map in $[\varphi]$. Hence, we have a continuous map $H : S^k \times [0,1] \to \mathcal{C}_p(X)$ such that $H|_{S^k \times \{0\}} = \varphi$ and $H|_{S^k \times \{1\}} = \varphi'$. Furthermore, this map can be chosen to be PL with respect to the triangulation of $\mathcal{C}_p(X)$. Therefore, by the same construction as above, we obtain that an integral current $c_H := (pr_2)_*(c(H))$. It is clear $\partial(c_H) = c_\varphi - c_{\varphi'}$ since the push-forward $(pr_2)_*$ commutes with the boundary map ∂.

$\square$

Alternatively, the restriction of φ to the interior of each top dimensional simplix Δ_j^k $(1 \le j \le N$, N is the number of top dimensional simplices) gives a map $\varphi : \Delta_j^k \to \Gamma_{n_j}$, where Δ_j^k is the j-th k-dimensional simplex and n_j is the maximum number such that Γ_{n_j} contains the image of $\varphi|_{\Delta_j^k}$. The piecewise linear property of φ with respect to the stratified structure of $\mathcal{C}_p(X)$ have the following property:

(∗) *For each $t \in \Delta_j^k$, $\varphi(s) = \sum_i a_i(s)V_i(s)$ with the property that $a_i(s) = a_i$ is constant in s and $V_i(s)$ is irreducible.*

For each j, $1 \le j \le N$, set $Z_{\Delta_j^k} := \sum_i a_i Z_{i,j}^k$, where $Z_{i,j}^k := \{(s,z) \in \Delta_j^k \times X | z \in V_i(s)\}$. It is clear that $Z_{\Delta_j^k}$ is an integral current. Therefore,

$$\phi_j^k := (pr_2)_*(Z_{\Delta_j^k}) \tag{6.3}$$

is then an integral current of real dimension $2p + k$, where $pr_2 : \Delta_j^k \times X \to X$ is the projection onto the second factor. Set $Z(\varphi) := \sum_{j=1}^N \phi_j^k$.

Lemma 6.3.2. *The closure of $Z(\varphi)$ is an integral cycle in X.*

Proof. Since φ is piecewise linear with respect to the triangulation of $\mathcal{C}_p(X)$. The image of φ on each $(k-1)$-dimensional simplex Δ_i^{k-1} is in Γ_{m_i}, where m_i is the maximum number such that Γ_{m_i} contains the image of $\varphi|_{\Delta_i^{k-1}}$. Each $\varphi|_{\Delta_i^{k-1}}$ defines a current $\phi_i^{k-1} := (pr_2)_*(Z_{\Delta_i^{k-1}})$ as in (6.3). The sum

$$\sum_i \phi_i^{k-1}$$

is zero since, for each ϕ_i^{k-1}, there is exactly one $\phi_{i'}^{k-1}$ such that they have the same support but different orientation.

$\square$

Let c_φ be the total $(k+2p)$-cycle in X determined by φ. We will simply use c instead of c_φ unless it arises confusion.

Remark 6.3.1. *c_φ, as current, has restricted type $c_\varphi = [c_\varphi]_{p+k,p} + [c_\varphi]_{p+k-1,p+1} + \cdots + [c_\varphi]_{p,p+k}$.*

If c is homologous to zero, we denote it by $c \sim_{hom} 0$, i.e., $[\varphi] \to 0$ in $H_{2p+k}(X, \mathbb{Z})$ under the natural transformation $L_p H_{2p+k}(X) \to H_{2p+k}(X, \mathbb{Z})$ (see, e.g., [L2], p.185). This condition translates into the fact that there exists an integral topological $(2p+k+1)$-chain $\tilde{c}$ such that $\partial \tilde{c} = c$.

We denote by $\mathrm{Map}(S^k, \mathcal{C}_p(X))$ the set of piecewise linear maps with respect to a triangulation of $\mathcal{C}_p(X)$ from the k-dimensional sphere to the abelian topological monoid $\mathcal{C}_p(X))$ of p-cycles.

Set

$$\mathrm{Map}(S^k, \mathcal{C}_p(X))_{hom} \subset \mathrm{Map}(S^k, \mathcal{C}_p(X))$$

the subset of such maps $\varphi : S^k \to \mathcal{C}_p(X)$ whose total cycles c_φ is homologous to zero in $H_{2p+k}(X, \mathbb{Z})$. There is a natural induced compact open topology on the space of such maps $\mathrm{Map}(S^k, \mathcal{C}_p(X))$ (see, e.g., Whitehead [Wh]).

Now $\mathcal{Z}_p(X)$ is the group completion of the topological monoid $\mathcal{C}_p(X)$ ([F1], [L1]). In the following, we will denote by $\mathrm{Map}(S^k, \mathcal{Z}_p(X))$ the set of piecewise linear maps with respect to a triangulation of $\mathcal{Z}_p(X)$ from the k-dimensional sphere to the abelian topological group $\mathcal{Z}_p(X)$ of p-cycles.

Let $\varphi : S^k \to \mathcal{Z}_p(X)$ be a PL map which is homotopic to zero. Hence there exists a map $\tilde{\varphi} : D^{k+1} \to \mathcal{Z}_p(X)$ such that $\tilde{\varphi}$ is PL with respect to a triangulation of $\mathcal{Z}_p(X)$ and $\tilde{\varphi}|_{S^k} = \varphi$. Then $\tilde{\varphi}$ determines an integral current, i.e., the total $(k+1+2p)$-chain $\tilde{c}$ such that the boundary of $\tilde{c}$ is c, i.e., $\partial \tilde{c} = c$. From the definition, we have $\varphi \in \mathrm{Map}(S^k, \mathcal{Z}_p(X))_{hom}$. Denote by $\mathrm{Map}(S^k, \mathcal{Z}_p(X))_0$ the subspace of $\mathrm{Map}(S^k, \mathcal{Z}_p(X))_{hom}$ consisting of elements φ which are homotopic to zero.

6.3.1 The generalized Abel-Jacobi map on $\mathrm{Map}(S^k, \mathcal{Z}_p(X))_{hom}$

In this subsection, suppose that $c \sim_{hom} 0$, i.e., $[\varphi] \to 0$ in $H_{2p+k}(X, \mathbb{Z})$ under the natural transformation $L_p H_{2p+k}(X) \to H_{2p+k}(X, \mathbb{Z})$ (see, e.g., [L2], p.185). This condition translates into the fact that there exists an integral topological $(2p+k+1)$-chain $\tilde{c}$ such that $\partial \tilde{c} = c$.

Consider

$$\omega \in \left\{ \bigoplus_{r \geq k+1, r+s=k+1} \mathcal{E}^{p+r,p+s} \right\}, \quad d\omega = 0$$

and we define

$$\Phi_\varphi(\omega) = \int_{\tilde{c}} \omega.$$

We claim:

Proposition 6.3.1. Φ_φ *is well-defined, i.e.,* $\Phi_\varphi(\omega)$, *as an element in*

$$\left\{ \bigoplus_{r,s \geq 0, r+s=k+1} H^{p+r,p+s}(X) \right\}^* \Big/ H_{2p+k+1}(X, \mathbb{Z}),$$

depends only on the cohomology class of ω. *Here we identify* $H_{2p+k+1}(X, \mathbb{Z})$ *with the image of the composition*

$$H_{2p+k+1}(X, \mathbb{Z}) \xrightarrow{\rho} H_{2p+k+1}(X, \mathbb{C}) \cong H^{2p+k+1}(X, \mathbb{C})^* \xrightarrow{\pi} \left\{ \bigoplus_{r,s \geq 0, r+s=k+1} H^{p+r,p+s}(X) \right\}^*, \quad (6.4)$$

where ρ *is the coefficient homomorphism and* π *is the projection onto the subspace.*

Proof. We need to show

1. For another choice of $\omega' \in \bigoplus_{r \geq k+1, r+s=k+1} \mathcal{E}^{p+r,p+s}$, $\omega - \omega' = d\alpha$, we have $\int_{\tilde{c}} \omega = \int_{\tilde{c}} \omega'$.

2. If $\tilde{c}'$ is another integral topological chain such that $\partial \tilde{c}' = c$, then we also have $\int_{\tilde{c}} \omega = \int_{\tilde{c}'} \omega$, where $\tilde{c}'$ is the currents determined by $\tilde{\varphi}$.

To show the part 1), note that we can choose α such that $\omega - \omega' = d\alpha$ for some α with $\alpha^{r,s} = 0$ if $r \leq k + p$ by the Hodge decomposition theorem for differential forms on X. Hence

$$\int_{\tilde{c}} \omega - \int_{\tilde{c}} \omega' = \int_{\tilde{c}} d\alpha = \int_c \alpha = 0$$

by the Stokes Theorem and the reason of type. This shows that the definition of Φ_φ is independent of the cohomology class of

$$[\omega] \in \left\{ \bigoplus_{r \geq k+1, r+s=k+1} H^{p+r,p+s}(X) \right\}.$$

To show the part 2), note that $\partial(\tilde{c} - \tilde{c}') = 0$ and hence $\tilde{c} - \tilde{c}' = \lambda$ is an integral topological cycle and hence $\int_\lambda$ lies in the image of the composition in (6.4). Hence $\int_\lambda$ is well-defined independently of the choice of $\tilde{c}$ such that $\partial\tilde{c} = c$, as an element in

$$\left\{ \bigoplus_{r,s \geq 0, r+s=k+1} H^{p+r,p+s}(X) \right\}^* \Big/ H_{2p+k+1}(X, \mathbb{Z}).$$

Hence we obtain a well-defined element

$$\Phi_\varphi \in \left\{ \bigoplus_{r \geq k+1, r+s=k+1} H^{p+r,p+s}(X) \right\}^* \Big/ H_{2p+k+1}(X, \mathbb{Z}).$$

This completes the proof of the Proposition.

$\square$

Therefore by Proposition 6.3.1 we have a well-defined homomorphism

$$\Phi : \mathrm{Map}(S^k, \mathcal{Z}_p(X))_{hom} \rightarrow \left\{ \bigoplus_{r \geq k+1, r+s=k+1} H^{p+r,p+s}(X) \right\}^* \Big/ H_{2p+k+1}(X, \mathbb{Z}) \qquad (6.5)$$

given by $\Phi(\varphi) = \Phi_\varphi$.

6.3.2 The restriction of Φ on $\mathrm{Map}(S^k, \mathcal{Z}_p(X))_0$

In this subsection, we will study the restriction of Φ in (6.5) to the subspace

$$\mathrm{Map}(S^k, \mathcal{Z}_p(X))_0 \subset \mathrm{Map}(S^k, \mathcal{Z}_p(X))_{hom},$$

i.e., all PL maps from S^k to $\mathcal{Z}_p(X)$ which are homotopic to zero. Note that the image of Φ is in

$$\{ \bigoplus_{r \geq k+1, r+s=k+1} H^{p+r,p+s}(X) \}^* / H_{2p+k+1}(X, \mathbb{Z}).$$

Let $\varphi : S^k \rightarrow \mathcal{Z}_p(X)$ be an element in $\mathrm{Map}(S^k, \mathcal{Z}_p(X))_0$. Denote by c the total $(k+2p)$-cycle (maybe degenerated) determined by φ. Hence there exists a map $\tilde{\varphi} : D^{k+1} \rightarrow \mathcal{Z}_p(X)$ such that $\tilde{\varphi}|_{S^k} = \varphi$ and the associated total $(k+1+2p)$-chain $\tilde{c}$ such that the boundary of $\tilde{c}$ is c, i.e., $\partial\tilde{c} = c$.

The restriction of the generalized Abel-Jacobi map Φ to the subspace of $\mathrm{Map}(S^k, \mathcal{Z}_p(X))_0$ is the map

$$\Phi_0 : \mathrm{Map}(S^k, \mathcal{Z}_p(X))_0 \to \left\{ \bigoplus_{r \geq k+1, r+s=k+1} H^{p+r,p+s}(X) \right\}^* \Big/ H_{2p+k+1}(X, \mathbb{Z}).$$

Now

$$\tilde{c} \in \left\{ \bigoplus_{r,s \geq 0, r+s=k+1} \mathcal{E}_{p+r,p+s} \right\}$$

and

$$c \in \left\{ \bigoplus_{r,s \geq 0, r+s=k} \mathcal{E}_{p+r,p+s} \right\}.$$

Hence

$$\Phi_\varphi(\omega) = \int_{\tilde{c}} \omega = 0$$

for $\omega \in \oplus_{r>k+1, r+s=k+1} \mathcal{E}^{p+r,p+s}$ with $d\omega = 0$ by the reason of type. Therefore $\Phi_\varphi = 0$ on $\oplus_{r>k+1, r+s=k+1} H^{p+r,p+s}(X)$. That is to say, the image of Φ on the subspace $\mathrm{Map}(S^k, \mathcal{Z}_p(X))_0$ is in

$$H^{p+k+1,p}(X)^* / \{ H^{p+k+1,p}(X)^* \cap \rho(H_{2p+k+1}(X, \mathbb{Z})) \}.$$

6.3.3   The reduction of Φ to $L_p H_{2p+k}(X)_{hom}$

We reduce the domain Φ to the quotient

$$\mathrm{Map}(S^k, \mathcal{Z}_p(X))_{hom} / \mathrm{Map}(S^k, \mathcal{Z}_p(X))_0 \cong \pi_0(\mathrm{Map}(S^k, \mathcal{Z}_p(X))_{hom}.$$

Now, if there are two maps $\varphi : S^k \to \mathcal{Z}_p(X)$ and $\varphi' : S^k \to \mathcal{Z}_p(X)$ such that φ is homotopic to φ'. Denote by c (resp. c') the total $(k+2p)$-cycle determined by φ (resp. φ'). For

$$\omega \in \left\{ \bigoplus_{r>k+1, r+s=k+1} \mathcal{E}^{p+r,p+s} \right\}, \quad d\omega = 0,$$

since $c - c' \sim_{hom} 0$, we have $\Phi_{\varphi-\varphi'}\omega = \Phi_\varphi \omega - \Phi_{\varphi'}\omega = 0$ and

$$\Phi_\varphi = \Phi_{\varphi'} \in \left\{ \bigoplus_{r>k+1, r+s=k+1} H^{p+r, p+s}(X) \right\}^* \Big/ H_{2p+k+1}(X, \mathbb{Z})$$

by the discuss in §3.2.

Therefore, we have a commutative diagram

$$\begin{array}{ccc}
\operatorname{Map}(S^k, \mathcal{Z}_p(X))_0 & \overset{i}{\hookrightarrow} & \operatorname{Map}(S^k, \mathcal{Z}_p(X))_{hom} \\
\downarrow \Phi_0 & & \downarrow \Phi \\
H^{p+k+1, p}(X)^*/H_{2p+k+1}(X, \mathbb{Z}) & \overset{i}{\hookrightarrow} & \left\{ \bigoplus_{r \geq k+1, r+s=k+1} H^{p+r, p+s}(X) \right\}^* \Big/ H_{2p+k+1}(X, \mathbb{Z}).
\end{array}$$

From this, we reduce Φ to a map

$$\Phi_{tr} : \pi_0(\operatorname{Map}(S^k, \mathcal{Z}_p(X))_{hom} \to \left\{ \bigoplus_{r>k+1, r+s=k+1} H^{p+r, p+s}(X) \right\}^* \Big/ H_{2p+k+1}(X, \mathbb{Z}) \qquad (6.6)$$

given by $\Phi_{tr}(\varphi) = \Phi_\varphi$. Here $/H_{2p+k+1}(X, \mathbb{Z})$ means modulo the image of the composition map

$$H_{2p+k+1}(X, \mathbb{Z}) \overset{\rho}{\to} H_{2p+k+1}(X, \mathbb{C}) = \left\{ \bigoplus_{r+s=2p+k+1} H^{r,s}(X) \right\}^* \to \left\{ \bigoplus_{r>k+1, r+s=k+1} H^{p+r, p+s}(X) \right\}^*.$$

We complete the construction of the generalized Abel-Jacobi map on homologically trivial part in Lawson homology

$$L_p H_{2p+k}(X)_{hom} := \pi_0(\operatorname{Map}(S^k, \mathcal{Z}_p(X))_{hom},$$

i.e., the kernel of the natural transformation $L_p H_{2p+k}(X) \to H_{2p+k}(X, \mathbb{Z})$.

Remark 6.3.2. *This map defined above is exactly the usual Abel-Jacobi given by Griffiths when $k = 0$ since there is a natural isomorphism $L_p H_{2p}(X)_{hom} \cong \operatorname{Griff}_p(X)$ (cf. [F1]). This map Φ on $L_0 H_k(X)_{hom}$ is trivial since $L_0 H_k(X)_{hom} = 0$ by Dold-Thom theorem (cf. [DT]).*

Remark 6.3.3. *Our generalized Abel-Jacobi map has been generalized to Lawson homology groups by the author. The range of the more generalized Abel-Jacobi map will be certain Deligne (co)homology. The tools used there are "sparks" and "differential characters" systematically studied by Harvey, Lawson and Zweck [HLZ] and [HL2].*

Remark 6.3.4. *Sometimes we also use* $AJ_X(c)$ *to denote* $\Phi_{tr}(\varphi)$*, where c is the cycle determined by* φ*.*

6.4 The non-triviality of the generalized Abel-Jacobi map

The natural question is the existence of smooth projective varieties such that the generalized Abel-Jacobi map Φ_{tr} on $L_p H_{2p+k}(X)_{hom}$ is non-trivial for both $p > 0$ and $k > 0$. The following example is a family of smooth 4-dimensional projective varieties X with $L_1 H_3(X)_{hom} \neq 0$, even modulo the torsion.

Example: Let E be an elliptic curve and Y be a smooth projective algebraic variety such that the Griffiths group of 1-cycles of Y tensored with $\mathbb{Q}$ is nontrivial. Set $X = E \times Y$. Let $[\omega] \in H^{4,0}(X)$ be a non zero element. By Künnuth formula, we have $[\omega] = [\alpha] \wedge [\beta]$ for some $0 \neq [\alpha] \in H^{1,0}(E)$ and $0 \neq [\beta] \in H^{3,0}(Y)$.

Let $\imath : S^1 \to E$ be a homeomorphism onto its image such that $\imath(S^1) \subset E$ is not homologous to zero in $H_1(E, \mathbb{Z})$. Let $\varphi : S^1 \to \mathcal{Z}_1(X)$ be a continuous map given by

$$\varphi(t) = (\imath(t), W) \in \mathcal{Z}_1(X), \qquad (6.7)$$

where $W \in \mathcal{Z}_1(Y)$ a fixed element such that W is homologous to zero but W is not algebraic equivalent to zero, i.e., $W \in \text{Griff}_1(Y)$. The existence of W is the assumption. Then there exists an integral topological chain U such that $\partial U = W$. Using the notation above, the cycle c determined by φ is $\imath(S^1) \times W$. Now $c = \imath(S^1) \times W$ is homologous to zero in X. Indeed,

$$\partial(\imath(S^1) \times U) = \partial(\imath(S^1)) \times U + (S^1) \times \partial U = \imath(S^1) \times W = c \qquad (6.8)$$

Hence $\partial \tilde{c} = \imath(S^1) \times U + \gamma + \partial(something)$, where $\partial \gamma = 0$. Therefore we have

$$\int_{\tilde{c}} \omega = \int_{\imath(S^1) \times U} \omega = \left(\int_{\imath(S^1)} \alpha \right) \left(\int_U \beta \right).$$

Proposition 6.4.1. *Suppose Y is a smooth threefold and* $W \in \mathcal{Z}_1(Y)$ *such that the image*

$AJ_Y(W)$ of W under the Griffiths' Abel-Jacobi map AJ_Y is non torsion in

$$H^{3,0}(Y)^*/\mathrm{Im}H_3(Y,\mathbb{Z}).$$

The map φ is given by (6.7) as above. Then the map $\Phi_{tr}(\varphi) \in H^{4,0}(X)/\mathrm{Im}H_4(X,\mathbb{Z})$ is nontrivial, even modulo torsion.

Proof. By Künneth formula, we have $H^{4,0}(E \times Y) \cong H^{1,0}(E) \otimes H^{3,0}(Y)$ and $H_4(E \times Y,\mathbb{Z}) \cong H_4(Y,\mathbb{Z}) \oplus \{H_1(E,\mathbb{Z}) \otimes H_3(Y,\mathbb{Z})\} \oplus \{H_2(E,\mathbb{Z}) \otimes H_2(Y,\mathbb{Z})\}$ modulo torsion. Let

$$\pi : H_4(E \times Y,\mathbb{Z}) \to \{H^{4,0}(E \times Y)\}^*$$

be the natural map given by $\pi(u)(\alpha \otimes \beta) = \int_u \alpha \wedge \beta$ for $u \in H_4(E \times Y,\mathbb{Z})$ and $\alpha \in H^{1,0}(E)$ and $\beta \in H^{3,0}(Y)$. Now $\pi(u) \neq 0$ only if $u \in H_1(E,\mathbb{Z}) \otimes H_3(Y,\mathbb{Z})$. Hence we get

$$\{H^{4,0}(E \times Y)\}^*/\mathrm{Im}H_4(E \times Y,\mathbb{Z}) \cong \{H^{1,0}(E)^* \otimes H^{3,0}(Y)^*\}/\mathrm{Im}\{H_1(E,\mathbb{Z}) \otimes H_3(Y,\mathbb{Z})\}.$$

Therefore, by the definition of generalized Abel-Jacobi map and (6.8), we have

$$AJ_Y(\imath(S^1) \times W)(\alpha \wedge \beta) = \Phi_{tr}(\varphi)(\alpha \wedge \beta)$$
$$= \left(\int_{\imath(S^1)} \alpha \right) \cdot \left(\int_U \beta \right)$$
$$= \left(\int_{\imath(S^1)} \alpha \right) \cdot (AJ_Y(W)(\beta))$$

i.e., $AJ_Y(\imath(S^1) \times W) = \int_{\imath(S^1)} \otimes AJ_Y(W)$.

Note that the map $\int_{\imath(S^1)} : H^{1,0}(E) \to \mathbb{C}$ is in the image of the embedding $H_1(E,\mathbb{Z}) \hookrightarrow H^{1,0}(E)^*$. But $AJ_Y(W)$ is a non-torsion element in $H^{3,0}(Y)^*/\mathrm{Im}H_3(Y,\mathbb{Z})$. Now the conclusion of the proposition is from the following lemma.

Lemma 6.4.1. Let V_m and V_n be two $\mathbb{C}$-vector spaces of dimension of m and n, respectively. Suppose that $\Lambda_m \subset V_m$, $\Lambda_n \subset V_n$ be two lattices, respectively. If $b \in V_n$ is a non torsion element in V_n, i.e., kb is not in Λ_n for any $k \in \mathbb{Z}^*$, then $a \otimes b$ is not in $\Lambda_m \otimes \Lambda_n$ for any $0 \neq a \in \Lambda_m$.

Proof. Set $\mathrm{rank}(\Lambda_m) = m_0$, $\mathrm{rank}(\Lambda_n) = n_0$. Let $\{e_i\}_{i=1}^{m_0}$, $\{f_j\}_{j=1}^{n_0}$ be two integral basis

89

of Λ_m, Λ_n, respectively. If the conclusion in the lemma fails, then

$$a \otimes b = \sum_{i=1}^{m_0} \sum_{j=1}^{n_0} k_{ij} e_i \otimes f_j,$$

for some $k_{ij} \in \mathbb{Z}$. By taking the conjugation, we can suppose that V_m and V_m are real vector spaces with lattices Λ_m and Λ_n, respectively.

Suppose that $a = \sum_{i=1}^{m_0} k_i e_i$, where $k_i \in \mathbb{Z}$, $i = 1, \cdots, m_0$ are not all zeros. The above formula reads as

$$\sum_{1}^{m_0} k_i e_i \otimes b = \sum_{i=1}^{m_0} \sum_{j=1}^{n_0} k_{ij} e_i \otimes f_j$$

i.e.,

$$\sum_{i=1}^{m_0} e_i \otimes (k_i b - \sum_{j=1}^{n_0} k_{ij} f_j) = 0.$$

Since $\{e_i\}_{i=1}^{m_0}$ is a basis in Λ_m and hence they are linearly independent over $\mathbb{R}$ in V_m, we get

$$k_i b - \sum_{j=1}^{n_0} k_{ij} f_j = 0$$

for any $i = 1, 2, \cdots, m_0$. By assumption, at least one of k_i is nonzero since a is nonzero vector in V_m. The last formula contracts to the assumption that kb is not in Λ_n for any $k \in \mathbb{Z}^*$. This completes the proof of the lemma and hence the proof of the proposition. $\square$

$\square$

More generally, we have the following Proposition

Proposition 6.4.2. *Suppose Y is a smooth threefold such that the image $AJ_Y(\mathrm{Griff}_1(Y))$ of $\mathrm{Griff}_1(Y)$ under the Griffiths' Abel-Jacobi map AJ_Y tensored by $\mathbb{Q}$ is are infinitely dimensional $\mathbb{Q}$-vector space over $\mathbb{Q}$ in*

$$\{H^{3,0}(Y)^* / \mathrm{Im} H_3(Y, \mathbb{Z})\} \otimes \mathbb{Q}.$$

For each $W \in \mathrm{Griff}_1(Y)$, The map φ_W is given by (6.7) as above. Then the image

$$\left\{ \Phi_{tr}(\varphi_W) | W \subset \mathrm{Griff}_1(Y) \right\} \otimes \mathbb{Q} \subset \left\{ H^{4,0}(X)/\mathrm{Im}H_4(X,\mathbb{Z}) \right\} \otimes \mathbb{Q}$$

is an infinite dimensional $\mathbb{Q}$-vector space.

Proof. We only need to show that:

(∗) Let $N > 0$ be an integer and $W_1, \cdots, W_N \in \mathrm{Griff}_1(Y)$ be N linearly independent elements under Griffiths Abel-Jacobi map. Then $\varphi_{W_1}, \cdots, \varphi_{W_N} \in L_1H_3(E \times Y)_{hom} \otimes \mathbb{Q}$ are linearly independent even under the generalized Abel-Jacobi map.

The claim (∗) follows easily from Proposition 6.4.1 above since if $\varphi_{W_1}, \cdots, \varphi_{W_N}$ are linearly dependent implies that $W_1, \cdots, W_N$ are linearly dependent by Proposition 6.4.1. This contradicts to the assumption. $\qquad\square$

Now for suitable choice of the 3-dimensional projective Y, for example, the general quintic hypersurface in $\mathbb{P}^4$ (cf. [G]) or the Jacobian of a general algebraic curve with genus 3 (cf. [Ce]) and the 1-cycle W whose image under Abel-Jacobi map is nonzero, in fact, it is infinitely generated for general quintic hypersurface in $\mathbb{P}^4$ (cf. [Cl]). Recall the definition of Abel-Jacobi map, $AJ_Y(W) = \int_U$ module lattice $H^3(Y,\mathbb{Z})$, we have $\int_U \beta \neq 0$ for this choice of W and some nonzero $[\beta] \in H^{3,0}(Y)$.

This example also gives an affirmative answer the following question:

Question: Can one show that $L_pH_{2p+j}(X)_{hom}$ is nontrivial or even infinitely generated for some projective variety X where $j > 0$?

Remark 6.4.1. *From the proof of the above propositions, we see that the non-triviality of Griffiths' Abel-Jacobi map on Y implies the non-triviality of the generalized Abel-Jacobi map on homologically trivial part of certain Lawson homology groups for X, i.e., all the Abel-Jacobi invariants can be found by generalized Abel-Jacobi map. In [Cl], Clemens showed the for general quintic 3-fold, the image of the Griffiths group under the Griffiths' Abel-Jacobi map can be infinitely generated, even modulo torsion.*

Remark 6.4.2. *Friedlander proved in [F2] the non-triviality of $L_rH_{2p}(X)_{hom}$ for certain complete intersections by using Nori's method in [N], which is totally different the construction here. There is no claim of any kind of infinite generated property of Lawson homology in his paper.*

Remark 6.4.3. *Nori [N] has generalized Theorem 6.1.1 and has shown that even the Griffiths' Abel-Jacobi map is trivial on some Griffiths group but the Griffiths group itself is nontrivial, even non torsion. By using a total different, explicit and elementary construction, the author has constructed singular rational 4-dimensional projective varieties*

X such that $L_1H_3(X)_{hom}$ is infinitely generated [H4]. But the Able-Jacobi map is not defined on singular projective variety (at least I don't know).

From the proof of Proposition 6.4.1, we observe that, for Y as above, and M is a projective manifold, if there is a map $i : S^k \to M$ such that

$$\int_{i(S^k)} : H^{k,0}(M) \to \mathbb{C}$$

is non-trivial as element in $\{H^{k,0}(M)\}^*$, then the value of the generalized Abel-Jacobi map Φ_{tr} at $\varphi : S^k \to \mathcal{Z}_1(X)$ defined by

$$\varphi(t) = (i(t), W) \in \mathcal{Z}_1(M \times Y)$$

is non-trivial, even modulo torsion.

Note that if the *complex Hurewicz homomorphism* $\rho \otimes \mathbb{C} : \pi_k(X) \otimes \mathbb{C} \to H_k(M, \mathbb{C})$ is surjective or even a little weaker condition, i.e., the composition

$$\pi_k(X) \otimes \mathbb{C} \to H_k(M, \mathbb{C}) \to \{H^{k,0}(M)\}^*$$

is surjective, we have the non-triviality of the map $\int_{i(S^k)} : H^{k,0}(M) \to \mathbb{C}$ if $H^{k,0}(M) \neq 0$. Here the map $\pi : H_k(M, \mathbb{C}) \to \{H^{k,0}(M)\}^*$ is the Poincaré duality the projection $H^k(M, \mathbb{C}) \to H^{k,0}(M)$ in Hodge decomposition.

As a direct application to the Main Theorem in [[DGMS], §6] and also Theorem 14 in [NT], we have the following result on higher dimensional hypersurface.

Proposition 6.4.3. *Let M be a smooth hypersurface in P^{n+1} and $n > 1$. Then the composition map*

$$\pi_k(X) \otimes \mathbb{C} \to H_k(M, \mathbb{C}) \to \{H^{k,0}(M)\}^*$$

is surjective for any simply connected Kähler manifolds.

Therefore we obtain the following result:

Theorem 6.4.1. *For any $k \geq 0$, there exist a projective manifold X of dimension $k + 3$ such that $L_1H_{k+2}(X)_{hom} \otimes \mathbb{Q}$ is nontrivial or even infinite dimensional over $\mathbb{Q}$.* □

By using the Projective Bundle Theorem in [FG], we get the following result:

Theorem 6.4.2. *For any $p > 0$ and $k \geq 0$, there is a smooth projective variety X such that $L_pH_{k+2p}(X)_{hom} \otimes \mathbb{Q}$ is infinite dimensional vector space over $\mathbb{Q}$.* □

Chapter 7

A homomorphism from Lawson homology to Deligne Cohomology

7.1 Introduction

In this chapter, all varieties are defined over $\mathbb{C}$. Let X be a projective variety with dimension n. The **Lawson homology** $L_pH_k(X)$ of p-cycles is defined by

$$L_pH_k(X) := \pi_{k-2p}(\mathcal{Z}_p(X)) \quad for \quad k \geq 2p \geq 0,$$

where $\mathcal{Z}_p(X)$ is provided with a natural topology such that it is an abelian topological group (cf. [F1], [L1]). For general background, the reader is referred to [L2].

In [FM], Friedlander and Mazur showed that there are natural maps, called **cycle class maps**

$$\Phi_{p,k} : L_pH_k(X) \to H_k(X)$$

from Lawson homology to singular homology.

Define

$$L_pH_k(X)_{hom} := \ker\{\Phi_{p,k} : L_pH_k(X) \to H_k(X)\}.$$

Temporarily suppose X is a complex manifold. Let Ω_X^k the sheaf of holomorphic k-form on X. The **Deligne complex of level p** is the complex of sheaves

$$\underline{\mathbb{Z}}_{\mathcal{D}}(p) : 0 \to \mathbb{Z} \overset{(2i\pi)^p}{\to} \Omega_X^0 \to \Omega_X^1 \to \Omega_X^2 \to \cdots \to \Omega_X^{p-1} \to 0$$

The **Deligne cohomology** of X in level p we mean the hypercohomology of this complex:

$$H_{\mathcal{D}}^*(X, \mathbb{Z}(p)) := \mathbb{H}^*(X, \underline{\mathbb{Z}}_{\mathcal{D}}(p)).$$

For more details on Deligne cohomology, the reader is referred to [EV].

Using the theory of differential characters introduced by Cheeger-Simons in [Ch] and [CS], systematically studied by Harvey, Lawson and Zweck in [HLZ] and the theory of D-bar sparks developed by Harvey and Lawson in [HL1] and [HL2], we can defined a natural homomorphism from the Lawson homology to the corresponding Deligne cohomology, which coincides to the generalized Abel-Jacobi map defined by the author in [H6]. The main result in this chapter is the following:

The main result in this chapter is the following

Theorem 7.1.1. *Let X be a smooth projective manifold with dimension n. We have a well-defined homomorphism*

$$\hat{a} : L_p H_{k+2p}(X) \rightarrow H_{\mathcal{D}}^{2(n-p)-k}(X, \mathbb{Z}(n-p-k-1)),$$

given by

$$\hat{a}([f]) = \widehat{a}_f$$

which coincides with the generalized Abel-Jacobi map defined in [H6] when $\hat{a}$ is restricted on $L_p H_k(X)_{hom}$ and the projection of the image of $\hat{a}$ under δ_2 is the natural map $\Phi_{p,k}$.

The notation $\widehat{a}_f$ and δ_2 will be defined below.

7.2 Sparks and differential characters

It might necessary to sketch the background materials of sparks and differential characters for our construction later. For the details of the materials used here, see [HLZ],[HL1] and [HL2]. First we recall some definitions we need. In this section X denotes smooth manifold unless otherwise noted.

Definition 7.2.1. *Set $\mathcal{E}^k(X) :=$ the space of smooth differential forms k-form on X with C^∞-topology; $\mathcal{D}^k(X) := \{\phi \in \mathcal{E}^k(X) \mid supp(\phi) \text{ is compact}\}$. We say the space of currents of degree k (and dimension $n - k$) on X, it means the topological dual space*

$\mathcal{D}'^k(X) \equiv \mathcal{D}'_{n-k}(X) := \{\mathcal{D}^{n-k}(X)\}'.$

$\mathcal{R}^k(X) :=$ *the locally rectifiable currents of degree k(dimension n − k) on X*

$\mathcal{IF}^k(X) :=$ *the locally integrally flat currents of degree k on X*

$\mathcal{I}^k(X) :=$ *the locally integral currents of degree k on X*

The following notation was firstly given in [HLZ]:

Definition 7.2.2. *The space of* sparks *of degree k on X is defined to be*

$$\mathcal{S}^k(X) := \{s \in \mathcal{D}'^k(X) | da = \phi - R \quad \text{where} \quad \phi \in \mathcal{E}^{k+1}(X) \quad \text{and} \quad R \in \mathcal{IF}^{k+1}(X)\}$$

Definition 7.2.3. *For each integer k, $0 \leq k \leq n$, we define the de Rham-Federer characters of degree k to be the quotient*

$$\widehat{\mathbb{H}}^k(X) := \mathcal{S}^k(X)/\{d\mathcal{D}'^{k-1}(X)\} + \mathcal{IF}^k(X)\}$$

The equivalence class in $\widehat{\mathbb{H}}^k(X)$ of a spark $a \in \mathcal{S}^k(X)$ will be denoted by $\hat{a}$.

It has been proved that ϕ and R in the decomposition of da above is unique [[HLZ], Lemma 1.3]. Moreover, there are two well-defined surjective maps:

$$\delta_1 : \widehat{\mathbb{H}}^k(X) \to \mathcal{Z}_0^{k+1}(X); \quad \delta_1(\hat{a}) = \phi$$

and

$$\delta_2 : \widehat{\mathbb{H}}^k(X) \to H^{k+1}(X, \mathbb{Z}); \quad \delta_2(\hat{a}) = [R],$$

where $\mathcal{Z}_0^{k+1}(X)$ denotes the lattice of smooth d-closed, degree $k + 1$ forms on X with integral periods.

Now we can give the definition of Riemannian Abel-Jacobi map. Let X be compact Riemannian manifold. Any current R on X, has a Hodge decomposition (cf. [HP])

$$R = H(R) + dd^*G(R) + d^*dG(R)$$

where H is harmonic projection and G is the Green operator. Also recall that d commutes with G, so that if R is a cycle, then $dG(R) = 0$. For $R \in \mathcal{IF}^{k+1}(X)$, set $a(R) := -d^*G(R)$ then

$$da(R) = H(R) - R$$

i.e., $a(R)$ is a *Hodge spark*. Let $\widehat{a(R)} \in \widehat{\mathbb{H}}^k(X)$ denote the differential character corresponding to the Hodge spark $a(R)$. Set

$$\text{Jac}^k(X) := H^k(X; \mathbb{R})/H^k_{\text{free}}(X; \mathbb{Z}); \quad \mathcal{B}^{k+1}(X) := d\mathcal{IF}^k(X)$$

then we have a well-defined map

$$\hat{a} : \mathcal{B}^{k+1}(X) \to \text{Jac}^k(X)$$

which is called the k-th *Riemannian Abel-Jacobi map*.

In [HL1], the concept of **homological spark complex** and its associated **group of homological spark classes** are given. In [HL2], a generalized version of homological spark complex is given as follows:

Definition 7.2.4. *A* **homological spark complex** *is a triple of cochain complexes* (F^*, E^*, I^*) *together with morphisms*

$$\Psi : I^* \to F^* \supset E^*$$

such that:

1. *$\Psi(I^k) \cap E^k = \{0\}$ for $k > 0$,*

2. *$H^*(E) \cong H^*(F)$, and*

3. *$\Psi(I^0) \to F^0$ is injective.*

Definition 7.2.5. *In a given spark complex (F^*, E^*, I^*) a* **spark of degree** k *is a pair*

$$(a, r) \in F^k \oplus I^{k+1}$$

which satisfies the spark equation

1. *$da = e - \Psi(r)$ for some $e \in E^{k+1}$, and*

2. $dr = 0$.

The group of sparks of degree k is denoted by $\mathcal{S}^k = \mathcal{S}^k(F^*, E^*, I^*)$.

Definition 7.2.6. *Two sparks $(a, r), (a', r') \in \mathcal{S}^k(F^*, E^*, I^*)$ are equivalent if there exists a pair*

$$(b, s) \in F^{k-1} \oplus I^k$$

such that

1. $a - a' = db + \Psi(s)$, and

2. $r - r' = -ds$.

*The set of equivalence classes is called the **group of spark classes of degree** k associated to the given spark complex and will be denoted by $\widehat{\mathbb{H}}^k(F^*, E^*, I^*)$ or simply $\widehat{\mathbb{H}}^k(F)$.*

As usual, let $Z^k(E) = \{e \in E^k | de = 0\}$ and set

$$Z_I^k(E) := \{e \in Z^k(E) | [e] = \Psi_*(\rho) \quad \text{for} \quad \text{some} \quad \rho \in H^k(I)\}$$

where $[e]$ denotes the class of e in $H^k(E)$ (note that $de = 0$). The following lemma was proved in [HLZ]:

Lemma 7.2.1. *There exist well-defined surjective homomorphisms:*

$$\delta_1 : \widehat{\mathbb{H}}^k(F) \to Z_I^k(E) \quad and \quad \delta_2 : \widehat{\mathbb{H}}^k(F) \to H^{k+1}(I)$$

given on any representing spark $(a, r) \in \mathcal{S}^k$ by

$$\delta_1(a, r) = e \quad and \quad \delta_2(a, r) = [r]$$

where $da = e - \Psi(r)$ as in definition 7.2.5.

Example: The following concrete example is the main object which will be dealt with in the next section. Now let X be a projective manifold with dimension n. Set

$$F^m = \mathcal{D}'^m(X, q) := \oplus_{r+s=m, r<q} \mathcal{D}'^{r,s}(X) \quad \text{and} \quad \bar{d} = \Psi \circ d$$

97

where

$$\Psi : \mathcal{D}'^m(X) \to \mathcal{D}'^m(X,q)$$

is the projection $\Psi(a) = a^{0,m} + a^{1,m-1} + \cdots + a^{q-1,m-q+1}$.

$$E^m = \mathcal{E}^m(X,q) := \oplus_{r+s=m,r<q} \mathcal{E}^{r,s}(X) \quad \text{and} \quad \bar{d} = \Psi \circ d$$

and

$$I^m = \mathcal{I}^m(X)$$

It has been shown in [[HL2],§14] that the above triple (F^*, E^*, I^*) is a homological spark complex . The group of associated spark classes in degree m will be denoted by $\widehat{\mathbb{H}}^m(X,q)$. To this homological spark complex, it has been that

$$\ker(\delta_1) = H_{\mathcal{D}}^{m+1}(X, \mathbb{Z}(q)).$$

7.3 The construction of the homomorphism from Lawson homology to Deligne cohomology

In this section, X denotes the projective manifold. Let $\mathcal{Z}_p(X)$ be the space of algebraic p-cycles with a natural topology (cf. [F1], [L1]) and a base point, i.e., the 'null' p-cycle 0 and let $\Omega^k \mathcal{Z}_p(X)$ be the loop space with the given base point. Explicitly, suppose $S^k = \mathbb{R}^k \cup \infty$. We have

$$\Omega^k \mathcal{Z}_p(X) = \{f : S^k \to \mathcal{Z}_p(X) | f \text{ is continuous with } f(\infty) = 0\}.$$

Given such a continuous map $f : S^k \to \mathcal{C}_p(X)$, we can find a map $g : S^k \to \mathcal{C}_p(X)$ such that g is homotopic to f and g is piecewise linear with regard to a triangulation of $\mathcal{C}_p(X)$ and hence one can define a current c_g over X and this current c_g is a cycle. Moreover c_g depends only on the homotopy class of g. For the detail of the construction, see [FL1] and [H6].

Consider the example from the last section with $m = 2(n-p)-k-1, q = n-p-k-1$. All the following argument will focus on this homological spark complex.

Definition 7.3.1. Set $a_f := -\Psi(d^*G(c_f))$. Then (a_f, c_f) is called the Hodge spark of the map $f : S^k \to \mathcal{Z}_p(X)$. Let $\hat{a}_f \in \widehat{\mathbb{H}}^{2(n-p)-k-1}(X, n-p-k-1)$ be the differential character corresponding to the Hodge spark (a_f, c_f).

Lemma 7.3.1. *If $f \in \Omega^k \mathcal{Z}_p(X)$, then $\delta_1(\widehat{a}_f) = 0 \in \widehat{\mathbb{H}}^{2(n-p)-k-1}(X, n-p-k-1)$.*

Proof. Let $H(c_f)$ be the harmonic part of c_f. Note that $H(c_f)$ and c_f are of the same $(*,*)$ type. Moreover,

$$c_f \in \oplus_{r+s=k, |r-s| \leq k} \mathcal{D}'_{p+r,p+s}(X).$$

By the type reason, the projection of $H(c_f)$ on

$$\oplus_{r+s=k, r \geq k+1} \mathcal{D}'_{p+r,p+s}(X)$$

is zero. This is exactly the image of $\widehat{a}_f$ under δ_1. $\qquad\square$

Hence we get an element $\widehat{a}_f \in \ker(\delta_1) = H_{\mathcal{D}}^{2(n-p)-k}(X, \mathbb{Z}(n-p-k-1))$. We denote this map by the notation $\widehat{a} : \Omega^k \mathcal{Z}_p(X) \to H_{\mathcal{D}}^{2(n-p)-k}(X, \mathbb{Z}(n-p-k-1))$, $\widehat{a}(f) = \widehat{a}_f$.

Remark 7.3.1. *By the argument above, we have actually defined an element*

$$\widehat{a}_f \in H_{\mathcal{D}}^{2(n-p)-k}(X, \mathbb{Z}(n-p-k)).$$

If the map f is contractible, then we have the following:

Lemma 7.3.2. *If $f : S^k \to \mathcal{Z}_p(X)$ is contractible, then $\widehat{a}(f) = 0$.*

Proof. Let $F : D^{k+1} \to \mathcal{Z}_p(X)$ be the extension of f, i.e., $F|_{\partial(D^{k+1})} = f$. Let c_F be the current over X defined by F. As showed in Lemma 7.3.1, c_F is a cycle. Moreover, it is a boundary, i.e., $c_F = \partial(c_f)$. By Corollary 12.11 in [HLZ], we have $\widehat{a}_f = H(\widehat{\Psi(c_F)})$. Since

$$c_F \in \left\{ \bigoplus_{r+s=k+1, |r-s| \leq k+1} \mathcal{D}'_{p+r,p+s}(X) \right\},$$

the projection of $H(c_F)$ under Ψ on

$$\bigoplus_{r+s=k, r > k+1} \mathcal{D}'_{p+r,p+s}(X)$$

is zero. Note that Ψ commutes with the Laplace operator, we have $\widehat{a}_f = 0$. $\qquad\square$

By the Lemma 7.3.1 and Lemma 7.3.2, we have a well-defined map

$$\hat{a} : L_p H_k(X) \to H_{\mathcal{D}}^{2(n-p)-k}(X, \mathbb{Z}(n-p-k-1)),$$

given by

$$\hat{a}([f]) = \widehat{a_f}.$$

Recall that the Deligne cohomology can be written as the middle part of an short exact sequence

$$0 \to \frac{H^{2(n-p)-k-1}(X, n-p-k-1)}{H^{2(n-p)-k-1}(X, \mathbb{Z})} \to H_{\mathcal{D}}^{2(n-p)-k}(X, \mathbb{Z}(n-p-k-1)) \to \ker(\Psi_*) \to 0,$$

where

$$H^{2(n-p)-k-1}(X, n-p-k-1) = \left\{ \bigoplus_{r+s=2(n-p)-k-1, r<n-p-k-1} H^{r,s}(X) \right\}$$

and

$$\ker(\Psi_*) = H^{2(n-p)-k}(X, \mathbb{Z}) \cap \left\{ \bigoplus_{r+s=2(n-p)-k, |r-s| \le k+2} H^{r,s}(X) \right\}$$

Proposition 7.3.1. *The restriction of the above map to $L_p H_k(X)_{hom} := \ker \{ \Phi_{p,k} : L_p H_k(X) \to H_k(X) \}$ is the generalized Abel-Jacobi map defined in [H6] if we identify the $H^{r,s}(X)$ with $\{H^{n-r,n-s}(X)\}^*$ for all $0 \le r, s \le n$ and $H^q(X, \mathbb{Z})$ with $H_{2n-q}(X, \mathbb{Z})$.*

  Proof. Recall the definition that

$$\Phi : L_p H_k(X)_{hom} \to \left\{ \bigoplus_{r>k+1, r+s=k+1} H^{p+r,p+s}(X) \right\}^* \Big/ H_{2p+k+1}(X, \mathbb{Z})$$

is given by $\Phi([f]) = \Phi_f$, where $\Phi_f(\omega) = \int_{\tilde{c}} \omega \pmod{H_{2p+k}(X, \mathbb{Z})}$ with $\partial(\tilde{c}) = c_f$. The Lemma 12.10 in [HLZ] implies that the two constructions coincide.

$\square$

Remark 7.3.2. *It is easy to see that the image of $\Phi_{p,k}$ is in $\ker(\Psi_*)$. Hence the natural map $\Phi_{p,k} : L_p H_k(X) \to H_k(X)\}$ factors through $\hat{a}$ and the map δ_2 in the first 3×3 grid given in [[HL2],§14].*

Remark 7.3.3. *Gillet and Soulé [GS] first showed that the Griffiths' intermediate Abel-Jacobi map coincides with the Riemannian Abel-Jacobi. Harris [Ha] also discussed some* *related topics.*

In summary, we have the following

Theorem 7.3.1. *Let X be a smooth projective manifold with dimension n. We have a well-defined homomorphism*

$$\hat{a} : L_p H_{k+2p}(X) \longrightarrow H_{\mathcal{D}}^{2(n-p)-k}(X, \mathbb{Z}(n-p-k-1)),$$

given by

$$\hat{a}([f]) = \widehat{a_f}$$

which coincides with the generalized Abel-Jacobi map defined in [H6] when $\hat{a}$ is restricted on $L_p H_k(X)_{hom}$; and the projection of the image of $\hat{a}$ under δ_2 is the natural map $\Phi_{p,k}$. □

Remark 7.3.4. *In general, the map $\hat{a}$ is a nontrivial homomorphism even if restricted on $L_p H_k(X)_{hom}$ in the case of $k = 2p$, which has been proved by Griffiths [G]. For the case that $k > 2p$, the author showed the nontriviality by examples in [H6].*

Bibliography

[AKMW] D. Abramovich; K. Karu; K. Matsuki and J. Włodarczyk, *Torification and factorization of birational maps.* J. Amer. Math. Soc. 15 (2002), no. 3, 531–572 (electronic).

[A] F. Almgren, *The homotopy groups of the integral cycle groups.* Topology 1 1962 257–299.

[AC] A. Albano and A. Collino, *On the Griffiths group of the cubic sevenfold.* Math. Ann. 299 (1994), no. 4, 715–726.

[AH] M. F. Atiyah and F. Hirzebruch, *Analytic cycles on complex manifolds.* Topology 1 1962 25–45.

[BCC] E. Ballico; F. Catanese and C. Ciliberto. *Classification of irregular varieties. Minimal models and abelian varieties.* Proceedings of the conference held in Trento, December 17–21, 1990. Lecture Notes in Mathematics, 1515. Springer-Verlag, Berlin, 1992. vi+149 pp. ISBN 3-540-55295-2

[BS] S. Bloch and V. Srinivas, *Remarks on correspondences and algebraic cycles.* Amer. J. Math. 105 (1983), no. 5, 1235–1253.

[Ce] G. Ceresa, *C is not algebraically equivalent to C^- in its Jacobian.* Ann. of Math. (2) 117 (1983), no. 2, 285–291.

[Ch] J. Cheeger, *Multiplication of differential characters.* Symposia Mathematica, Vol. XI (Convegno di Geometria, INDAM, Rome, 1972), pp. 441–445. Academic Press, London, 1973.

[CS] J. Cheeger and J. Simons *Differential characters and geometric invariants.* Geometry and topology (College Park, Md., 1983/84), 50–80, Lecture Notes in Math., 1167, Springer, Berlin, 1985.

[Cl] H. Clemens, *Homological equivalence, modulo algebraic equivalence, is not finitely generated.* Inst. Hautes Études Sci. Publ. Math. No. 58 (1983), 19–38 (1984).

[DGMS] P. Deligne; P. Griffiths; J. Morgan and D. Sullivan, *Real homotopy theory of Kähler manifolds*. Invent. Math. 29 (1975), no. 3, 245–274.

[DT] A. Dold and R. Thom, *Quasifaserungen und unendliche symmetrische Produkte*. (German) Ann. of Math. (2) 67 1958 239–281.

[EV] Hélène Esnault; Eckart Viehweg, *Deligne-Beĭlinson cohomology*. Beĭlinson's conjectures on special values of *L*-functions, 43–91, Perspect. Math., 4, Academic Press, Boston, MA, 1988.

[F1] E. Friedlander, *Algebraic cycles, Chow varieties, and Lawson homology*. Compositio Math. 77 (1991), no. 1, 55–93.

[F2] E. Friedlander, *Filtrations on algebraic cycles and homology*. Ann. Sci. École Norm. Sup. (4) 28 (1995), no. 3, 317–343.

[F3] Friedlander, Eric M., *Relative Chow correspondences and the Griffiths group*. (English. English, French summary) Ann. Inst. Fourier (Grenoble) 50 (2000), no. 4, 1073–1098.

[FG] E. Friedlander and O. Gabber, *Cycle spaces and intersection theory. Topological methods in modern mathematics* (Stony Brook, NY, 1991), 325–370, Publish or Perish, Houston, TX, 1993.

[FHW] E. Friedlander; C. Haesemeyer and M. Walker, *Techniques, computations, and conjectures for semi-topological K-theory* Preprint.

[FL1] E. Friedlander and B. Lawson, *A theory of algebraic cocycles*. Ann. of Math. (2) 136 (1992), no. 2, 361–428.

[FL2] E. Friedlander and B. Lawson, *Duality relating spaces of algebraic cocycles and cycles*. Topology 36 (1997), no. 2, 533–565

[FL3] E. Friedlander and B. H. Lawson, *Graph mappings and poincaré duality*, preprint.

[FM] E. Friedlander and B. Mazur, *Filtrations on the homology of algebraic varieties. With an appendix by Daniel Quillen*. Mem. Amer. Math. Soc. 110 (1994), no. 529, x+110 pp.

[GS] H. Gillet and C. Soulé ,*Arithmetic Chow groups and differential characters*. Algebraic *K*-theory: connections with geometry and topology (Lake Louise, AB, 1987), 29–68, NATO Adv. Sci. Inst. Ser. C Math. Phys. Sci., 279, Kluwer Acad. Publ., Dordrecht, 1989.

[G]   P. Griffiths, *On the periods of certain rational integrals I, II.* Ann. of Math. (2) 90(1969), 460-495; ibid. (2) 90(1969) 496–541.

[GH]   P. Griffiths and J. Harris, *Principles of algebraic geometry.* Reprint of the 1978 original. Wiley Classics Library. John Wiley & Sons, Inc., New York, 1994. xiv+813 pp. ISBN 0-471-05059-8

[Gro]   A. Grothendieck, *Hodge's general conjecture is false for trivial reasons.* Topology 8 1969 299–303.

[Ha]   B. Harris, *Differential characters and the Abel-Jacobi map.* Algebraic K-theory: connections with geometry and topology (Lake Louise, AB, 1987), 69–86, NATO Adv. Sci. Inst. Ser. C Math. Phys. Sci., 279, Kluwer Acad. Publ., Dordrecht, 1989.

[HLZ]   R. Harvey; B. Lawson and J. Zweck, *The de Rham-Federer theory of differential characters and character duality.* Amer. J. Math. 125 (2003), no. 4, 791–847

[HP]   R. Harvey and J. Polking, *Fundamental solutions in complex analysis. I and II.* The Cauchy-Riemann operator. Duke Math. J. 46 (1979), no. 2, 253–340.

[HL1]   R. Harvey and B. Lawson, *From Sparks to Grundles - Differential Characters* arXiv.org:math.DG/0306193

[HL2]   R. Harvey and B. Lawson, *D-bar sparks, I.* arXiv.org:math.DG/0512247

[Hi1]   H. Hironaka, *Resolution of singularities of an algebraic variety over a field of characteristic zero. I, II.* Ann. of Math. (2) 79 (1964), 109–203; ibid. (2) 79 1964 205–326.

[Hi2]   H. Hironaka, *Triangulations of algebraic sets,* Proc. of Symposia in Pure Math 29 (1975), 165-185.

[H1]   W. Hu, *Birational invariants defined by Lawson homology.* arXiv.org:math.AG/0511722.

[H2]   W. Hu, *The Generalized Hodge conjecture for 1-cycles and codimension two algebraic cycles.* arXiv.org:math.AG/0511725.

[H3]   W. Hu, *Some relations between the topological and geometric filtration for smooth projective varieties.* arXiv.org:math.AG/0603203.

[H4]   W. Hu, *Infinitely generated Lawson homology groups on some rational projective varieties.* arXiv.org:math.AG/0602517

[H5]   W. Hu, *A note on Lawson homology for smooth varieties with small Chow groups.* arXiv.org:math.AG/0602516

[H6]   W. Hu, *Generalized Abel-Jacobi map on Lawson homology.* In this thesis.

[H7]   W. Hu, *A map between Lawson homology and Deligne Cohomology.* In this thesis.

[J]   Uwe Jannsen, *Motivic sheaves and filtrations on Chow groups.* Motives (Seattle, WA, 1991), 245–302, Proc. Sympos. Pure Math., 55, Part 1, Amer. Math. Soc., Providence, RI, 1994.

[KMM]   J. Kollár; Y. Miyaoka and S.Mori, *Rationally connected varieties.* J. Algebraic Geom. 1 (1992), no. 3, 429–448.

[Lat]   R. Laterveer, *Algebraic varieties with small Chow groups.* J. Math. Kyoto Univ. 38 (1998), no. 4, 673–694.

[L1]   H. B. Lawson, Jr., *Algebraic cycles and homotopy theory.*, Ann. of Math. **129**(1989), 253-291.

[L2]   H. B. Lawson, Jr., *Spaces of algebraic cycles.* pp. 137-213 in Surveys in Differential Geometry, 1995 vol.2, International Press, 1995.

[Lew1]   J. D. Lewis, *A survey of the Hodge conjecture. (English. English summary) Second edition. Appendix B by B. Brent Gordon.* CRM Monograph Series, 10. American Mathematical Society, Providence, RI, 1999. xvi+368 pp. ISBN 0-8218-0568-1

[Lew2]   J. D. Lewis, *Three lectures on the Hodge conjecture.* (English. English summary) *Transcendental aspects of algebraic cycles,* 199–234, London Math. Soc. Lecture Note Ser., 313, Cambridge Univ. Press, Cambridge, 2004.

[Lieb]   D. Lieberman, *Numerical and homological equivalence of algebraic cycles on Hodge manifolds.* Amer. J. Math. 90 1968 366–374.

[Li1]   P. Lima-Filho, *Lawson homology for quasiprojective varieties.* Compositio Math. 84(1992), no. 1, 1–23.

[Li2]   P. Lima-Filho, *On the generalized cycle map.* (English. English summary) J. Differential Geom. 38 (1993), no. 1, 105–129

[NT]   J. Neisendorfer and L. Taylor, *Dolbeault homotopy theory.* Trans. Amer. Math. Soc. 245 (1978), 183–210.

[N]   M. Nori, *Algebraic cycles and Hodge-theoretic connectivity.* Invent. Math. 111 (1993), no. 2, 349–373.

[Pa]   K. H. Paranjape, *Cohomological and cycle-theoretic connectivity*. Ann. of Math. (2) 139 (1994), no. 3, 641–660.

[Pe]   C. Peters, *Lawson homology for varieties with small Chow groups and the induced filtration on the Griffiths groups*. Math. Z. 234 (2000), no. 2, 209–223.

[Ro]   A. A. Roĭtman, Rational equivalence of zero-dimensional cycles. (Russian) Mat. Sb. (N.S.) 89(131) (1972), 569–585, 671

[S]   C. Schoen, *On Hodge structures and nonrepresentability of Chow groups*. Compositio Math. 88 (1993), no. 3, 285–316.

[Vn]   Mircea Voineagu, *Semi-topological K-theory for certain projective varieties*. arXiv.org:math.KT/0601008

[V1]   C. Voisin, *Hodge theory and complex algebraic geometry. I*. Translated from the French original by Leila Schneps. Cambridge Studies in Advanced Mathematics, 76. Cambridge University Press, Cambridge, 2002. x+322 pp. ISBN 0-521-80260-1

[V2]   C. Voisin, *Hodge theory and complex algebraic geometry. II*. Translated from the French by Leila Schneps. Cambridge Studies in Advanced Mathematics, 77. Cambridge University Press, Cambridge, 2003. x+351 pp. ISBN 0-521-80283-0

[Wa]   M. E. Walker, *The morphic Abel-Jacobi map*.

[Wh]   G. W. Whitehead, *Elements of homotopy theory*. Graduate Texts in Mathematics, 61. Springer-Verlag, New York-Berlin, 1978. xxi+744 pp. ISBN 0-387-90336-4

[Wl]   J. Włodarczyk, *Toroidal varieties and the weak factorization theorem*. Invent. Math. 154 (2003), no. 2, 223–331.

www.ingramcontent.com/pod-product-compliance
Lightning Source LLC
Chambersburg PA
CBHW052016230326
41598CB00078B/3489